Successful Innovation

NEW HORIZONS IN THE ECONOMICS OF INNOVATION

General Editor: Christopher Freeman, *Emeritus Professor of Science Policy, SPRU – Science and Technology Policy Research, University of Sussex, UK*

Technical innovation is vital to the competitive performance of firms and of nations and for the sustained growth of the world economy. The economics of innovation is an area that has expanded dramatically in recent years and this major series, edited by one of the most distinguished scholars in the field, contributes to the debate and advances in research in this most important area.

The main emphasis is on the development and application of new ideas. The series provides a forum for original research in technology, innovation systems and management, industrial organisation, technological collaboration, knowledge and innovation, research and development, evolutionary theory and industrial strategy. International in its approach, the series includes some of the best theoretical and empirical work from both well-established researchers and the new generation of scholars.

Titles in the series include:

Successful Innovation

Towards a New Theory for the Management of
Small and Medium-sized Enterprises

Jan Cobbenhagen

*Programme Leader, MERIT and Managing Director of M3
New Business Creation at the University of Maastricht, The
Netherlands*

NEW HORIZONS IN THE ECONOMICS OF INNOVATION

Edward Elgar
Cheltenham, UK • Northampton, MA, USA

Published by
Edward Elgar Publishing Limited
Glensanda House
Montpellier Parade
Cheltenham
Glos GL50 1UA
UK

Edward Elgar Publishing, Inc.
136 West Street
Suite 202
Northampton
Massachusetts 01060
USA

A catalogue record for this book
is available from the British Library

Library of Congress Cataloguing in Publication Data

Cobbenhagen, Jan.
Successful innovation: towards a new theory for the management of small and medium-sized enterprises / Jan Cobbenhagen.
(New horizons in the economics of innovation)
Includes index.
1. Technological innovations–Management 2. Small business–Management. I. Title. II. Series.

HD45 C525 2000
658.5'14–dc21   00–024325

ISBN 1 84064 388 9

Printed by MPG Books Ltd, Bodmin, Cornwall

Contents

Figures

Tables

Boxes

Preface

Ten years ago, I began a voyage of discovery in what is known as the world of academia. As a PhD trainee, I was free to pursue my own path for four years. My goal at the time was to write a thesis and then, most probably, emigrate to that other world, the world of business. But I became fascinated by the academic world. Too fascinated, perhaps. I had soon gone beyond the borders of my homeland, the original doctoral study. Halfway through I even emigrated to another country, where I was given the opportunity to participate in a large-scale research project offering a much wider perspective than my doctoral study at the time. This experience was good for me. I came into contact with other research cultures and wanted to learn from them also. I soon found that I could apply the knowledge I had acquired in other projects as well. Gradually I became a sort of nomad, roaming from project to project. I have made many friends on my travels, and now and then an enemy. All of them gave me some spiritual luggage to take with me. Strangely enough, this only helped me to travel more lightly.

However, the goal with which I began my journey, the thesis, was steadily fading away. What I didn't realise was that the thesis was becoming more and more of an entry visa for the many countries I wanted to visit. Fortunately, Professor Friso den Hertog, my thesis supervisor and travel agent for this sometimes rather rough journey, regularly called my attention to this. He gave me invaluable support. Many ideas in this thesis spring from long, smoke-filled discussions in the creative and stimulating atmosphere of his office. But I first wanted to explore some more new territory, namely that of regional innovation policy. Subsequently, I finally found the drive and the peace to attain the goal I had set for myself at the journey's beginning. The result is before you. After a journey full of detours, roadblocks and side trips, my thesis has finally been completed. I now feel so at home in the world of academia that I intend to continue to explore it for some time. The original goal has now been attained, but new goals are already appearing on the horizon.

The preface is the proper place to thank my travelling companions for their support. First and foremost, I have to thank Friso den Hertog, my coach and supervisor, who has become a close personal friend. I am also thankful to my colleagues Edward Huizenga, Guido Philips and Ed van Sluijs, who read and commented on earlier reports on which this thesis is based. In addition, I would like to thank Professor Hans Pennings (University of Pennsylvania) for his support in earlier phases of the research project, especially with respect to the statistical analysis. Much of the nitty-gritty work accompanying this large research project was done by our research assistants, Wynand Bodewes, Annerose Buchel gen. van Steenbergen, Michiel Radder and Sandra Theelen, and secretaries Corien Gijsbers and Silvana de Sanctis. And I would like to thank all of them for their efforts. Furthermore, I am indebted to three anonymous reviewers of the *Academy of Management Journal* who provided me with invaluable feedback on some of the core issues of this thesis. Besides these professional companions, I should also like to thank my parents, who have given me the opportunity to study and follow my own path. I am proud of them for the way they have accepted that there will not be a third generation of hotel-owners in the family.

If writing a thesis means neglecting your family it is not worth the effort. Having said that, I must admit that it does take a heavy toll of free time which could be spent otherwise. For her continuous support and encouragement, I would like to thank my wife, Belinda. I dedicate this book to her and to my sons Roy, Tim† and Tom who, each in their own unique way, have shown me the value of life, but unfortunately also its fragility. We should stop and think of this more often in our busy lives.

Jan Cobbenhagen   Maastricht, March 4th, 2000

PART I

Theory and methodology

1. Practical framework

Usually, innovation is thought of in terms of technologies that give rise to new products or improvements in existing products. However, product and process innovation is not the primary bottleneck to progress. The bottleneck is management. US firms lag most in the management innovation required to take full advantage of their technology leadership. (Ray Stata, Chairman of Analog Devices; Stata, 1992)

1.1 INNOVATE

Companies are increasingly faced with intensifying competitive pressures. Customers pose ever more stringent demands regarding uniqueness, customisation, speed of delivery, quality, environmental performance and so on. In order to ensure their competitiveness, and even survival, companies must be able to meet these challenges by providing a continuous stream of new and improved products, processes and services. Innovation is not only of importance for a limited group of high-tech, manufacturing or large-scale companies. On the contrary, the need to innovate is universal, irrespective of size, sector or technological sophistication. In an environment where technologies, competitive positions and customer demands can change almost overnight and the life-cycles of products and services are getting shorter, the capacity to manage innovation successfully is crucial for the competitive power of a company. It is therefore no surprise that managing the innovative function of firms has gained increasing attention in both the business and academic communities.

In this chapter we will explain why innovation is more important today than ever before. Furthermore, we will show why organisational issues and management play a crucial role in mastering this elusive phenomenon called 'innovation'.

3

1.1.1 External challenges: innovation is necessary and important

Firms are faced with increasing (and increasingly globalising) competition. Market segments are fragmenting and demands for customised deliveries are forcing companies to produce smaller batches of a growing number of differentiated products. Customers are becoming ever more discerning and are imposing apparently conflicting demands on firms: customised solutions, fast delivery, high quality *and* sound environmental performance. Companies rise to the challenge and live up to these demands by delivering the products and services their customers want, with more features, better performance and appropriate levels of cost and quality. This implies a constant stream of new and improved products, processes and services, which is achieved through innovation. And this is true for all industries and service sectors, since the need for innovation is universal, with no particular relation to 'high-tech' firms or to the type, size or age of the company concerned.

For small and medium-sized enterprises (SMEs), which operate in an environment of greater uncertainty than larger firms, innovation can be a means of overcoming the uncertainty associated with being a price-taker and with a limited customer and product base (Storey, 1994). Of course, it can be argued that SMEs play a different role in innovation. But despite the glamorous role which Schumpeter (1934) ascribed to small firms (that of initiating 'gales of creative destruction'), we have to conclude that in general their role is less glamorous (Storey, 1994, p. 11). The much more conventional role which small firms play in innovation relates to the niches they might occupy. It is their ability to provide something marginally different in their products or services which distinguishes them from the more standardised products or services provided by the larger companies. Furthermore, the average SME is much less likely to conduct in-house research and development activities than a large company, let alone have a high proportion of its staff concerned exclusively with fundamental research. Even so, it is argued that SMEs are more likely to introduce fundamentally new innovations than larger companies (Pavitt et al., 1987). But this is the case for only a small group of SMEs, the high-tech or knowledge-intensive SMEs which often operate at the forefront of technical knowledge. The probability that an individual SME will introduce a product new to the world is of course much lower than the probability that a large company will do so. Nevertheless, although the potential to innovate differs between SMEs and large companies, the necessity to innovate in order to achieve a competitive advantage is equally important to both large and small firms.

And there is much to gain from innovation. Innovative firms tend to have larger market shares and higher growth rates and profits than non-innovative firms (Geroski and Machin, 1992). Furthermore, a link has been demonstrated between increasing research and development (R&D) expenditures and subsequent growth of turnover relative to those of competitors (Franko, 1989; Morbey, 1988). Successful process innovations usually have a positive influence on the productivity and efficiency of a firm. And technological renewal is generally considered to be a driving force behind economic growth and as such a remedy for structural economic downturns. Technological renewal is assumed to create new possibilities for companies, regions, sectors and countries (Ven, 1986). Thus innovation is an important precondition for economic growth and the improvement of national competitive positions. In fact, as Prahalad (1990) concludes from the dramatic shift in relative world leadership positions of some of the largest US companies: 'The nature of competition is changing toward new product development'. Good quality and low cost are no longer sufficient for global competitiveness. Specifically, it is their ability to develop innovative new products which allows some firms to grow at a faster rate than others; especially if they manage to introduce these new products more efficiently and at lower costs.

Research in the early 1980s (Booz, Allen and Hamilton, 1985; Sommerlatte, 1986) indicated that US managers believed that strengthening the innovative quality of enterprises would be the most important management challenge for the 1990s. And they have been proven right. Innovation in products, processes and services has indeed become a high-priority issue for firms in many manufacturing and service sectors. However, the attention devoted by firms to product development was subject to cycles in past decades. The heightened attention given by management to product creation in the early 1980s had to give way to increasing concentration on cost reduction in the late 1980s and early 1990s. But efforts aimed at increasing the productivity of firms can be negated if they are used in conjunction with outdated or obsolete technologies. On the other hand, devoting a lot of effort to R&D without having a proper organisation to reap the benefits from the technological advantage is also unlikely to lead to competitive strength. Coriat (1995), for instance, argues that while Europe is not behind in terms of basic research or R&D, it is losing ground in the process of developing from scientific discovery and invention to innovation and from innovation to the market. He therefore argues that, parallel to efforts within the framework of macroeconomic policy, policy makers should initiate specific long-term actions to foster a programme of organisational reform aimed at promoting organisational innovation.

Thus the need to devote attention to innovation has regained its high position on the manager's agenda. And policy makers are becoming increasingly aware of the importance of stimulating organisational knowledge of innovation in addition to technology development or transfer.

Innovation (or more specifically, renewal of products, processes and organisations) is not a fad but an essential activity for the long-term survival of organisations. Every product or process is subject to a life-cycle. Although these cycles vary in shape or length (depending on the product, process or line of business), sooner or later every product will become obsolete and call for a successor, an innovation. 'Innovation' as an activity, however, is nothing new. It is something humans have been engaged in ever since they invented the first stone tools and put them to use. What has changed is the pace at which firms have to innovate; and the inherent demands posed on the management of firms to master innovation efficiently and effectively.

For one thing, the life-cycles of products and technologies are becoming increasingly shorter (Olshavsky, 1980; Qualls et al., 1981; Sanderson and Uzumeri, 1990). Technology and market changes are imposing themselves more and more rapidly. Being first in the market with the right product often gives a competitive edge, since the 'newness' of a product is an important selling point. As a result, time has become one of the most important competitive weapons. Time savings allow companies to shorten their economic cycles, innovate more rapidly, and produce, sell and distribute more efficiently (Stalk, 1988; Uttal, 1987). The ability to shorten new product development processes thus becomes ever more important. In the electronic memory chip sector, for instance, it has been shown that a market introduction delay of six months will reduce profits by 30 per cent for products with a life-cycle of five years and 50 per cent for products with a three-year life-cycle (Gerpott and Wittkemper, 1991). Rapid new product development and timely introduction help a company to sustain its competitive position in a market and enable it to learn about the markets available to it. As a result of the ever shortening life-cycles of products and technologies, new products will find themselves quickly in the maturity phase, where competition is often fierce.

It can be argued that the industrialisation of the world has been characterised by continuous attention to company operating efficiency. A vast range of organisational disciplines, from operational research to human resources management, have both contributed to and benefited from this process (Hertog and Eijnatten, 1990). For a long time, these efforts were primarily aimed at the production process. Efficiency was much less a driving force for changes in the upstream trajectory, that is, in the process of developing products and processes. R&D was very much looked upon as a 'strategy of hope', based on the belief that results would eventually be achieved simply by bringing bright people together and supplying them with sufficient resources (Roussel et al., 1991). Efficiency was hardly regarded as a critical success factor in R&D. In the 1980s, however, we witnessed a fundamental shift in favour of the amount of attention devoted (in research as well as in management) to efficiency in company research and development activities (Cobbenhagen and Hertog, 1992). Increasing globalisation brought with it the threats of booming countries like Japan, and in later stages Korea and Taiwan, who were able to produce better products with more features at lower prices. Efficiency has now become an additional critical R&D success factor. In the current competitive environment, organisations realise that they not only have to become (more) innovative, they must become efficiently and effectively innovative as well:

> R&D managers are learning to make do in straitened circumstances. Continuing a trend begun in the late 1980s, they're ditching marginal projects, decentralising R&D efforts, creating teams to squire product ideas from lab to market, and figuring out how to collaborate fruitfully with outside experts. (*Business Week*, 1994b, p. 44)

As an R&D department has to be efficient as well as creative, management is confronted with the problem of balancing between 'control' and 'creative freedom'. On the one hand, a certain amount of creative freedom is needed to come up with something new. But on the other hand, something new does have to come out, so some control is needed. Furthermore, the success of a process is to a large extent determined by the choices that are made upstream. In the end there has to be a market for the new product. So there is no longer time for total creative freedom; new products have to be brought out, and they have to be brought out fast.

An increasing diversity of high-quality products must be brought to the market at an increasingly faster pace, preferably at lower prices. This poses high demands on the efficiency of the development process. Furthermore, it focuses attention on the consequences of choices made in the development process. In fact, efficiency in the plant is as much determined by the choices made in the course of product and process design as it is by optimisation schemes in the plant itself (Bolwijn et al., 1986).

Innovation can be manifest as a destructive as well as a constructive process and bring great uncertainty in its wake. After all, new products pose threats to existing products. Not only can investments in hardware diminish in value, but knowledge and experience are also subject to obsolescence. New processes sometimes result in a loss of jobs or redundancy for less highly skilled workers. But while innovation is indeed a risky process, refusing to innovate is if anything even more perilous, for then the threat may come from anywhere. Yet an investment in technological innovation alone offers no guarantee that a company will survive the competition. The relation between R&D expenditures and growth in profits, for example, appears to be quite weak (Morbey, 1988). Successful innovation thus means more than merely investing in R&D. It also means more than innovativeness. Among other things, it implies good coordination with production and with the market. Ideally, firms would like to bring a dominant design on to the market at just the right time, but this is no easy task. Although it pays to be the first in the market, success can rebound if the company's haste leads to design mistakes. Being too fast can turn out to be just as disastrous as being too slow (Box 1.1). Thus, not only innovation itself, but the effectiveness of the innovation process, has been gaining increasing importance as an area of competitive strength.

BOX 1.1 THE RIGHT TIMING...

A company concentrates on a certain process or product innovation, invests a good deal of capital in it, but appears to have bet on the wrong horse (for example, design). The company had made its choice too early. This scenario may take place primarily in capital-intensive companies with large R&D budgets. An example of this is the Philips Video 2000 system.

A company may also adopt a wait-and-see attitude before committing itself to a specific design or market. Delay or postponement, however, may also cause the company to miss the boat. This happened to NCR in the early 1970s when a line of newly designed cash registers with a total value of $140 million became unsaleable due to the obsolete technology on which it was based (Foster, 1986a). NCR built electromechanical cash registers which could not compete with the new, more economically produced and easier to operate electronic equipment.

1.1.2   Internal challenges: innovation is risky and expensive

Innovation bears not only the seeds of success but also those of destruction. It is a way to survive, but it can also be an easy way to get into financial trouble, for opportunities are created by taking risks. Companies can readily overstrain themselves in achieving their innovative ambitions. The adventures, ten years ago, of Philips in the area of mega chips and those of Gist-brocades and Shell in biotechnology are clear illustrations of this. Developing new products swallows funds and energy, especially when the firm moves outside its core activities. But even when new products and processes lead to immediate success on the market, innovations can still show their destructive side: new processes and products may cannibalise existing ones. This not only costs money but may lead to the loss of knowledge as well. And technological renewal is not necessarily without negative side effects. For one thing, the company runs the risk of being taken to task by society with respect to ecological, safety and health risks and its responsibilities as a provider of jobs. Partly due to the abundant and overwhelming influence of TV, radio and the press, reactions to innovations can be extremely fierce, as Unilever, for example, experienced after its introduction of 'OMO Power' (Box 1.2).

BOX 1.2 WHEN STAKES ARE HIGH... A FIERCE REACTION MIGHT OCCUR

Unilever's top brand detergents, Persil (U.K.) and OMO, were very gradually losing market share in Europe. Its main competitor, US-based Procter and Gamble (P&G) had even become the European leader in detergents. In April 1994, Unilever launched an attack to regain market leadership. With a marketing budget of over 500 million guilders, It introduced a new detergent, based on a 'new' manganese technology, which had the capacity to remove really difficult stains at low temperatures. This new detergent was marketed under the old brand names but with the adjective 'Power' added: OMO Power and Persil Power.

The technology was not really new, however. Unilever's main rival, P&G, had known about it for twenty years. It had experimented with it, but had never used it in its products, mainly because detergents based on this new technology were too strong, weakening clothing after several washings. Unilever improved on this technology and managed to produce a detergent which did not weaken clothes, provided it was used at the right temperature.

But information about the destructive capacity of this new detergent enabled P&G to launch a ferocious and very successful counterattack. A lot was at stake, and the battle became very fierce indeed. P&G publicly announced that independent testing had proven that Unilever's new detergent damaged clothing. It provided the press with reports of independent lab tests along with the clothing used in the tests. Each package contained two similar garments: one which was said to have been washed in the new detergent and which was totally ruined and a similar one, in mint condition, which was said to have been washed in P&G's own detergent, Ariel. When independent tests by consumer groups showed similar results (weakened fibres after about 15 washes at high temperature), consumers lost faith.

Unilever used full-page advertisements to reassure customers and gave them a full guarantee. In addition, it lowered the concentration of the new catalyst. But the damage had already been done. The credibility of the strong brand names OMO and Persil had been damaged and Unilever will have a tough time recouping the massive investments in this new detergent (about 300 million guilders on R&D and 500 million guilders on marketing).

Sources: *Business Week* (1994a); FEM (1994) and an interview with Mr R.J. Duggan (Director of the Crosfield Group, Unilever)

Finally, increasing globalisation is constantly redefining market shares and competitive relations. Managers are thus faced with the task of innovating in an ever-changing environment. This makes it hard to establish realistic goals. Innovation can therefore not be managed as if it were a steady-state process.

1.2 HARNESSING THE BEAST

Innovation is thus Janus-faced: it is a way to survive, but it is also an easy way to get into financial trouble, since opportunities are created by taking risks. This raises the question as to *how* innovative success can be achieved and sustained.

1.2.1 Research and development

For one thing, a firm has to be *innovative* in order to be *successfully innovative*. And being innovative implies more than just having creative people in the organisation: it also means being able to stimulate, nurture and exploit the creativity slumbering within every organisation. One important pool of creativity to tap into is research and development.

The question as to how innovative success can be achieved and sustained is not easy to answer. In many firms, the tension between opportunities and risks is reflected by quantitative reactions, by initiating and terminating investments in research and development. In consequence, a traditionally cogent question for managers is: 'Are we investing enough in the renewal of products and processes to keep up with the competition?'. When companies in many Dutch manufacturing sectors compared their R&D investments with that of their national or international competitors, the result was an increase in R&D investments in the early 1980s. This increase was mainly based on technological and scientific arguments, that is, the risk of lagging behind in the global race to develop new and promising fields such as biotechnology and the use of new materials. The larger enterprises, in particular, optimistically embarked on large-scale, organisationally complex co-operative relationships.

In the late 1980s, however, the tide changed and the dangerous and destructive underside of innovation became evident (Cobbenhagen and Hertog, 1995). Margins and profits were eroded by the rising costs of innovation and stagnating economic growth. Large-scale technological co-operation programmes in the fields of microelectronics, information technology, biotechnology and new materials were launched with high ambitions but did not always deliver the hoped-for success. Some newly opened plants were even threatened with closure and many new products were cannibalising existing cash cows. The increased pace of product innovation was leaving companies breathless. Many firms put on the brakes. R&D budgets decreased and the responsibility for innovation policies was brought closer to the market, that is, to divisions and business units rather than corporate headquarters. In the Netherlands, the net result of this move was a shrinking of business expenditure on R&D (as a percentage of GDP) from 1.4 per cent in 1987 to 1.0 per cent in 1991 and a slimming of R&D staff (Ministry of Economic Affairs, 1994). The reasoning behind this was that companies had become involved with too many organisationally and commercially complex adventures in fields both technologically and scientifically far from their areas of strength. In many large firms (for example; Philips and DSM), the concentration on 'core competencies' became the leading principle (Prahalad and Hamel, 1990).

Over the past few years, a changing of the tide can be observed again. Anxiety regarding the technological position of Dutch commerce and industry is again increasing. The arguments brought to the forefront resemble those used ten years ago. The concentration on core competencies is not only seen as a process of streamlining, but also as a way of creating new markets. Within the larger companies, this discussion relates particularly to the role of technology management at corporate level. While increasing emphasis on innovation in business is seen as a good cause, it is also recognised that it calls for renewed attention to be paid to the need for a long-term technological portfolio strategy.

This pendulum movement can be regarded as a quantitative reaction to crises. At moment 'A', the opportunities are taken as starting points and the inputs in the system are increased. At moment 'B', when outputs remain or fall below expectations, the tap is closed one turn. When this situation persists for too long, it is quite predictable that warnings will be uttered that the firm risks lagging behind unless inputs are increased quickly. This quantitative approach can also be recognised in the recent discussions about tax advantages and other financial support for innovative firms (Cobbenhagen et al., 1994a, 1994b).

The question remains, however, as to whether this quantitative approach alone is sufficient and whether better performance can only be achieved by spending more money on R&D. Examples from other European countries show that high R&D spending and/or good research is not a sufficient condition for innovative success. R&D must be translated into products sold on the market. But this is where many companies lose the battle. As Cresson and Bangemann (EC Commissioners) state in their Green Paper on Innovation:

> There is no doubt that Europe is among the leaders in terms of scientific achievement. But when it comes to commercial performance, our position in many high-technology sectors has deteriorated and continues to do so. This is what is sometimes called the 'European paradox' – we are good at research, but not at transforming these skills into a competitive advantage. (Cresson and Bangemann, 1995)

So success is not just a matter of investing in R&D and new equipment and employing high-quality personnel. If assessed wrongly, such measures may even bring the organisation into serious difficulties. And neither is success just a matter of 'being in the right market at the right time'. If that were the case, management could do little to turn a mediocre new product project into a success, since the project outcomes (according to this view) would be determined to a large extent by extraneous factors. Such fatalistic views result in rigidity and inertia and often lead to 'paralysis by analysis' (Cooper, 1986), with management becoming preoccupied with project selection. But financial projections provide managers with a false sense of security. Even sophisticated analytical instruments appear to add only limited value to decision making (Kaplan, 1986; Hertog and Roberts, 1992 and Roberts, 1993).

In successful innovative firms, the question is increasingly raised as to whether market, technical and financial information is sufficient. In these firms, the focus is far more on day-to-day management of the project. Instead of asking themselves: 'Is this the right project?', their question is: 'How do we go about turning this project into a winner?' (Cooper, 1986). And an important precondition for success is to have an organisation that is fit for this task; an organisation which has the technical, marketing and management capabilities effectively and efficiently to bring projects to successful conclusions. It is very noticeable, for example, that innovative companies succeed even in times of recession (DTI and CBI, 1993). Evidently, companies can improve despite the climate in which they may be operating.

Furthermore, R&D does not represent the only pool of creativity from which innovations may arise. Although many of the radical changes in past centuries were driven by technological breakthroughs, many others (albeit often less drastic) originated in the market. The challenge in these 'market-pull' innovations consists in the search for the best technology, either in-house or external, and the process of mutual adaptation between market and technology. The debate as to whether 'technology push' or 'market pull' is the more important source of product innovation is to a certain extent artificial, since the overwhelming evidence indicates that product innovations arise from the interaction of these two forces.

Innovation is therefore not the exclusive domain of the R&D department, but something which concerns the whole organisation. And innovative success is not just determined by R&D, but throughout the whole innovation chain, from the conception of an idea to the first product sold on the market and beyond. Innovation requires as much an inside-out attitude as an outside-in approach. In fact, the combination of market demand and technological capability has proved to be a good breeding ground for many successful innovations. If only the 'trick' were that simple.

Research has shown that the performance criteria on which large firms compete have been accumulating over time (Bolwijn and Kumpe, 1989) (see Figure 1.1). Starting with efficiency aspects which can be effectively controlled and verified, each subsequent factor becomes less controllable and verifiable. The more controllable competitive factors such as efficiency, customer orientation, quality and service appear to have lost their competitive value as more and more firms are successful in bringing them under control. The factor which currently has the highest competitive value is the least controllable of them all: innovativeness. It is therefore an illusion to think that routinisation, planning and other techniques which applied to the previous criteria still satisfy. Management has to search for new ways (new routines) to achieve innovative success. It is being confronted with a new or different type of problem that requires a different type of answer.

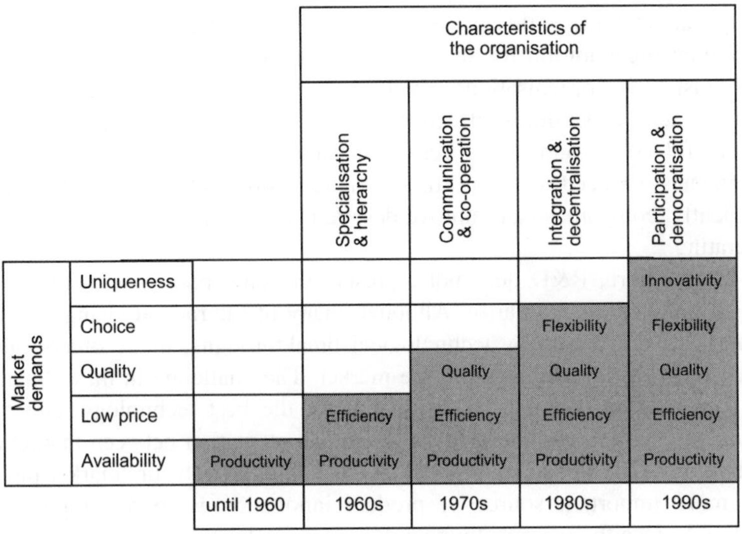

		Characteristics of the organisation			
		Specialisation & hierarchy	Communication & co-operation	Integration & decentralisation	Participation & democratisation
Market demands	Uniqueness				Innovativity
	Choice			Flexibility	Flexibility
	Quality		Quality	Quality	Quality
	Low price	Efficiency	Efficiency	Efficiency	Efficiency
	Availability	Productivity	Productivity	Productivity	Productivity	Productivity
		until 1960	1960s	1970s	1980s	1990s

Source: Bolwijn and Kumpe (1989)

Figure 1.1 Performance criteria

Prior to the 1980s, time and money did not play such an important role in managing innovation since the need for an effective organisation of innovation processes was less apparent. Innovation was regarded as being relatively separate from the routine processes and the customer. In managing their innovative efforts, firms tended to use models and instruments originally designed for the production environment.

1.2.2 Innovation is not a steady-state process

But although innovation is very much on the way to becoming a permanent activity of all firms ('The only constant factor we have around here is change', as one manager expressed it), it is not a steady-state process in the classical sense. Rather, it is a complex, non-routine process that confronts the organisation with dilemmas and uncertainties which are mostly unknown to production processes. To illustrate this, I should like to introduce a metaphor to help indicate the uncertainties and risks involved: the innovation process as a computer adventure game (see Box 1.3).[1]

BOX 1.3 AN ADVENTURE GAME AS A METAPHOR FOR INNOVATION PROCESSES

Computer adventures are a popular category of computer games which have a lot in common with innovation processes. In a computer adventure, the player has to find a treasure in a maze (often a castle or dungeon) by means of making simple decisions (choice of direction, actions to take or not to take, deciding which items to take and which not, and so on). On the way through the maze, the player will be cornered by various sinister events requiring quick and effective reactions. A critical situation that is assessed wrongly may result in the 'death' of the player (a new product failure). A player has a rough idea of the rules and of situations which might possibly be dangerous, but will never know everything exactly. During the trip through the maze, the player will be allowed to retain a few things (compare this with R&D projects) that are necessary for survival. However, it is not possible take everything, so critical choices have to be made (R&D project selection). Chances and uncertainties play an important role in the course of the game. To reduce the risks and insecurities it is possible to ask questions and gather information. But not all the information obtained is useful; some is even dysfunctional.

The player will learn by trial and error. Often the game is lost before the treasure (a successful new product) is found, but with the knowledge and experience gathered in playing the game (organisational learning), the chance to win a subsequent round has been increased significantly. However, if a player decides to play a different adventure (become engaged in a new line of activity), then the accumulated knowledge is significantly less useful.

In managing innovations, organisations face a completely different control problem than in managing steady-state processes like production or logistics. The difference between controlling an innovation process and controlling a steady-state production process reveals itself with respect to:

- *The time dimension:* like a production process, an innovation process has a beginning and an end, but the transitory nature of the innovation process makes it impossible to build in permanent facilities. Innovation processes ('the production of one new product') generally run much longer and are more stochastic than production processes ('the production of a known product'). Especially with science-based innovations, we see that the trajectory from 'research' to market introduction can sometimes take up to 20 years, with 'on' and 'off' periods of activity;

- *The system boundaries:* in a production process, people work in groups whose composition rarely changes. In the case of innovation processes, however, system boundaries are continuously changing or blurring as the composition of the group of people working on the innovation projects or involved from the outside (customers, suppliers) changes both during the process and from innovation to innovation as well;

- *The amount of routinisation:* contrary to the case in steady-state processes, the material and information flows in innovation processes are unique for each process. Routinisation (or learning by doing) occurs when knowledge and skills learnt in a particular process are reapplied to the same process. This 'gliding down the learning curve' which occurs in steady-state processes is difficult to achieve in innovation processes, since such processes all differ from each other. In production processes, one learns from the process with the aim of mastering the same process more effectively, whereas in innovation processes one must learn from the process in order to master future, similar or related processes more effectively. In the latter case, learning from experience occurs mainly in the course of different (innovation) processes and not within the same repeating process. This implies a different type or learning and routinisation;

- *The amount of uncertainty:* the degree of freedom in an innovation process is usually much higher than in a production process, especially at the start, when there is often only a vague idea about the characteristics and appearance of the new product or simply a list of specifications. During the process, the degree of freedom will decrease. Among other things, this uncertainty about the final outcome implies that managers cannot always function on the basis of existing norms and values, because these very norms and values are themselves subject to change and may no longer meet the requirements (Box 1.4).

BOX 1.4 RESINS: FROM ALCHEMY TO CHEMISTRY

ResChem (actual company but fictitious name) produces resins for paints and lacquers. The production or 'cooking' of resins, as it is called, is regarded by many as basically a handicraft process in which it is primarily experience that counts and not scientific (chemical) knowledge. When working with natural raw materials, one can expect surprises due to contamination (variances in the characteristics of the raw materials). These surprises can be tackled more effectively with intuition and experience than with scientific analysis or standard procedures. At least that is how ResChem has been doing it for years. The production managers jestingly use the term 'alchemy' to describe the company's production culture.

In the past ten years, we have witnessed ever-increasing progress in the fields of process technology and chemistry in the resins industry. Success in the market place is increasingly determined by the ability to combine contrasting customer requirements in a single product. This puts high demands on the controllability of the production processes. The development of production technology (better measurements and fewer tests) provides increasingly better answers. The interface between development and the plant is thus becoming of increasing importance.

This transition from high-grade craftsmanship to high-grade scientific and technological knowledge is not without ripples. As new young academics are brought in, a new culture is emerging, a culture of science which conflicts with the 'alchemical' culture of the lab technicians and production workers. The latter become easily annoyed when the young chemists' knowledge and experience surpass their own and they are sceptical about their scientific contributions. It is simply very difficult for them to accept that values and norms which led to success in the past are no longer satisfactory. For the young chemists, the situation is just as difficult. They enter the company with their heads full of knowledge and two left hands. They have a hard time trying to convince others that the existing values and norms are becoming outdated.

1.2.3 Organisational solutions

Since it is not the idea which leads to profits, but the whole innovation process, the question is not how much or how little a firm should innovate but rather whether it innovates effectively. The challenge then is to trace *how* the inputs in the innovation system are being transferred into the desired outputs. Illustrative in this respect is the quote by Stata (CEO of Analog Devices) at the beginning of this chapter. According to him, the real bottleneck to progress is not product and process innovation, but management; more specifically, a lack of management innovation. The question concerning the effectiveness of the organisational chain of products, processes and services can only be addressed when the black box of the innovating organisation is opened. Only then is it possible to determine how inputs can maximally be transferred into outputs. Or 'how ideas can be managed into good currency' (Ven, 1986). This implies a change in thinking from quantity towards quality: the quality of the innovating firm. Many firms have a lot to gain by changing the way innovation processes are managed, as Box 1.5 illustrates.

BOX 1.5 A LOT CAN BE GAINED...

There is a lot to gain by changing the way innovation processes are managed, as is clearly shown by the results achieved by the following companies that have completed break-out style changes in their design, development and manufacturing processes:

- Dowty Electronics reduced the time to market for a new modem from 24 to 8 months.
- Digital Equipment Corporation reduced the design cycle time for a new terminal by 73 per cent with part cost savings of 72 per cent and a 43 per cent reduction in part count.
- British Aerospace Commercial Aircraft Division reduced the manufacturing cycle for an aircraft from 63 weeks in 1987 to 35 weeks by the end of 1990.

Source: PA Consulting Group (1991)

Companies are therefore considering 'organisational solutions' to this new challenge: internal and external co-operation has increased and greater emphasis has been placed on interdisciplinary and holistic perspectives. With technology getting increasingly complex, no single company can do it all alone. Innovation is becoming more and more expensive. The cost of R&D for firms charting new territory is rising dramatically. And as product life-cycles become shorter, the time to generate enough profit to pay for the investment in R&D is also becoming shorter. Major steps can often only be made in collaboration with partners. Firms increasingly share knowledge by co-operating with customers, suppliers and even competitors, not only to share costs, but to gain a competitive advantage as well (Moss Kanter, 1989b). Knowledge transfer with other companies and knowledge institutions has become more important than ever before. This goes for both large and small companies. Knowledge transfer may limit itself to information exchange, but it may also assume more formalised structures. Knowledge development is therefore increasingly sought outside the firm. In particular, small firms, which often have limited or no research and/or development facilities at their disposal, look for knowledge outside. The keywords for them are 'networking' and 'trust' (Biemans, 1992; Docter et al., 1989). Typical examples of the new forms of networking organisations we see emerging are provided by the small and medium-sized enterprises (SMEs) which engage in virtual organisational relationships such as the KIC (Knowledge Intensive Industrial Clustering) projects around Océ, the copier producer. Océ hands down a well-defined product development task to a network of SMEs with complementary competencies and leaves the co-ordination to the network itself. The interesting development is that these SMEs get so well attuned to each other that some KIC networks remain effective after the job for Océ is done and even promote their network as if it were a firm.

1.3 ACHIEVING INNOVATIVE SUCCESS: ORGANISATION MATTERS

The risks and costs involved, as well as the complexity and uncertainties, make 'innovation' a difficult process to manage. Furthermore, global trends are setting increasingly stringent demands with respect to the effectiveness and efficiency of a firm's innovative efforts. The increasing number of product variations that firms have to produce to satisfy customers put greater demands on the flexibility of engineering, production and logistics. To be a successful innovator, more is needed than technical innovativeness or a clear market focus. The message that is learned from many new product failures is that organisation matters. The key to innovative success is an effective and efficient organisation of the firm's innovation function.

Numerous studies have shown the inability of once very successful firms or industrial sectors to cope with new requirements. The frequently cited problems of US industry less than a decade ago (rigidity, low quality and manufacturing problems) indicated that what was once the stereotype of industrial might had to struggle to keep up with the Asian countries. And the situation is not much different in Europe. The EC Commissioners Cresson and Bangemann (1995) state in their Green Paper on Innovation that one of the weaknesses of European innovation systems is the inadequate level of organisational innovation. In the Green Paper, it is argued that the use of innovation and technology management techniques gives the firms concerned an undeniable competitive advantage. The paper continues by providing a list of these tools, including such diverse items as quality approach, participative management, value analysis, design, economic intelligence, just-in-time production, re-engineering, performance ratings, and so on. But what we can basically distil from this is that it pays to invest in the organisation as well as in technological developments: provided the firm makes the right investments. This book is about the investments firms can and should make in their organisation in order to become more successful in their innovative attempts.

The more successful firms succeed by embracing new approaches, new technology and, above all, more efficient practices to get new products to the market fast (Jelinek and Bird Schoonhoven, 1990, p. 2) and this often requires complete break-out style changes in their design, development and manufacturing processes (PA Consultants, 1991). The specific organisational character of innovation trajectories therefore calls for a revision of the organisational paradigm. Many studies have since highlighted aspects of this new organisational paradigm (Nelson and Winter, 1982; Quinn, 1985, 1988; Takeuchi and Nonaka, 1986; Ven et al., 1989; Cooper, 1983; Hertog et al., 1996). This book aims to contribute to these efforts, in a modest way, by suggesting some ideas which could form part of such a paradigm shift.

Although this paradigm shift has gained wide attention in academic research, many firms, especially SMEs in more traditional industries, still manage innovation in practice as if it were a steady-state process. Organisation is hardly considered as a variable in the achievement of innovative success. Many firms are overly obsessed by technological breakthroughs and market research results. And when the results are disappointing, the money tap is turned off again. Such firms seem to ignore the value of a comprehensive organisation which can ensure that a product concept is developed into an actual product sold to the customers. A lot can go wrong in this process in terms of speed, costs, quality and focus, and having a technologically superior product does not provide enough ammunition to win the battle. This is the harsh lesson which many large companies have had to learn in recent years. Examples are abundant: Philips' Video 2000, Coca Cola's New Coke, IBM's PCjr, Dupont's Corfam and, more recently, Unilever's OMO Power. However, it is too easy to blame such mistakes on an excessive focus on pushing technology and a neglect of the market. In many of these cases, extensive market research indicated a serious market for the product. The idea for the new Coke actually originated in (observations of) the market.

But the battle is often lost due to a chocked-up bureaucratic organisation that impedes the firm from exploiting fully its technological or marketing competencies. The front runner who breaks away from the pack with innovative products, services and processes will also be endowed with organisational competence, the capacity to make the right connections at the right times and to get the most out of the company's human potential. The ease with which companies turn the money tap on and off for new technology and product promotion is often in sharp contrast to their lack of willingness to set the organisation on a radically new course. We can observe this not only in large manufacturing companies such as Philips, IBM, Volkswagen and Unilever, but also in smaller firms and in companies in the services sector. Simply turning the money tap on and off forms no basis for increasing effectiveness. The key to that is found in the manner in which the innovation function is organised, in the quality of the organisation.

1.4 NEW CONDITIONS OF COMPANY ORGANISATION

The growth of the large conglomerates in the 1970s was a result of diversification strategies. Building on and supported by a number of well-managed, frequently functionally orientated business activities, the large conglomerates were able to manage successfully a wide range of activities under a single umbrella. Business activities such as marketing, human resource management (HRM) and product development were mastered by more and more companies. Based on the increasing numbers of studies of such processes and the experiences of the pioneers, it became possible for the smaller firms to streamline their own functional processes more effectively as well. The management of company processes became professionalised. But as these capacities increasingly became common property, a situation arose in which the conglomerates had little advantage over less diversified firms. On the contrary, they were faced with the threat of losing their leading positions now that specialised knowledge was becoming a more important competitive weapon than knowledge about managing functional business processes, which had in any case become available to all. Companies which specialised acquired the advantage of superior knowledge and left the conglomerates behind (Willems and van den Wildenberg, 1993). Moreover, the disadvantages of a rigid functional organisation were becoming more and more evident (Hertog, 1979; Sitter, 1987): inflexibility, low motivation, co-ordination problems, sluggish innovation processes, and so on. To stay competitive, conglomerates took drastic measures, often in the form of major disinvestments. At the beginning of the 1980s, we thus saw a massive 'back-to-the-core' movement. Business units not directly related to the core business were sold, made independent or closed down. Companies whose core business turned out to be too narrow (partly due to macrodevelopments in areas such as transportation, energy, life-style and technology) resorted to the hybrid structure of a holding with business units (Willems and van den Wildenberg, 1993). This structure combines the features of a conglomerate of the 1970s with that of a specialised company of the 1980s. Connecting factors between business units could be found in raw materials, technology and the market, but coherence was not always evident. By the early 1990s, it had become clear that the connecting factor between business units could represent a significant competitive strength. Prahalad and Hamel (1990) use the term 'core competencies' to describe these central, strategic capabilities deemed crucial for the success of the company, which they define as 'the collective learning process in the organization, especially as to how to co-ordinate diverse production skills and integrate multiple streams of technology'. We shall come back to this concept in Chapter 2.

Parallel to the changes in the overall structure of the firm, we have also witnessed some dramatic shifts in the way innovation processes are organised. Thoughts and practices regarding the management of innovation processes evolved from narrow R&D management, sequential approaches through parallel flows to paradox management (Hertog, et al., 1996). To cope with the conflicting demands placed on firms (speed, creativity and flexibility on the one hand, formalisation, standardisation and quality, liability, ecological and safety regulations on the other), concerted action between core functions (such as marketing, R&D, mechanisation and manufacturing) at the business unit level is essential. This requires an integrated approach and draws attention to the challenge of overcoming the barriers between functional areas. In this respect, the emphasis is increasingly shifting from innovation as an R&D activity towards innovation as an integral process requiring the involvement of many company activities. In every phase, one has to anticipate what needs to be done in a later phase. However, as communication patterns become ever more complicated, the functional barriers will cause ever more problems. The cultural barriers between R&D and other functional departments found in many organisations may become significant adverse factors.

Business Week (22 October, 1993, p. 35) defined managing innovations as 'a riddle that requires a company to understand not just its markets and customers, but itself'. More than just a definition, this quotation reflects the feelings of many managers involved with innovations. Managing innovations does not only mean developing products that meet the real needs of the customers, but also nurturing and developing what a company is best at in various aspects: technology, marketing, logistics, service and management. And this complexity makes the process elusive and hard to manage. Examples of failures (such as Ford's Edsel, Dupont's Corfam, IBM's PCjr and Coca Cola's New Coke) have shown us that even top-performing companies goof up once in a while: as one would expect, for companies that always play it safe will not make much progress. Yet there is still a difference between companies that are constant front-runners and companies that have a little success now and then. It is the difference between firms that are successfully innovative and firms which have successful innovations. The real challenge is to stay successful for a long time. It is the difference between winning the 'Tour de France' for five consecutive years, like Miguel Indurain, and winning a single stage. Every firm covets the yellow sweater, but most just win a few stages (Box 1.6). But there are no 'born' front-runners, neither among cyclists nor among firms. Starting on the same terms as the rest of the pack, they gain competitive advantages by developing and maintaining superior skills and competencies over their rivals. The question that occupies such firms has to do with identifying the skills and competencies needed to acquire that competitive edge. What distinguishes the front-runners from the rest of pack? Many firms are searching for an answer to that question; it is also the basic question this book wishes to address.

This study is not alone in that attempt. Numerous studies focusing on the success factors of innovation processes can be found in the literature. In cycling terms, these are studies which identify the winners of a single stage. Whether small or large, firms encounter difficulties in building or preserving an innovative capability, that is, the skills to consistently generate viable new products or services. Some firms have been known to produce a successful innovation (for example, SmithKline Beecham's drug Tagamet, Philips's audiocassette player, or GM's Saturn) but encounter hurdles when they seek to generalise a single success into other product lines. In many firms, a successful innovation is a rare occasion and resembles a flash in the pan rather than something arising out of a coherent pattern of events. Other firms, however, appear to maintain permanent flows of viable new products and services. Each year, *Fortune* magazine publishes a list of 'most admired companies', in which firms like Rubbermaid and 3M figure prominently. A major reason for admiring such firms is their innovative capacity. Presumably, these firms can respond more adaptively and competitively to internal and external conditions.

BOX 1.6 FRONT-RUNNERS

PARIS, 24 July 1995 (Reuter) – Miguel Indurain of Spain completed an historic fifth consecutive Tour de France triumph on Sunday, taking his place as the greatest champion in the most demanding event in sport. Miguel Indurain's reign as king of the Tour de France may be set to run and run, according to the riders whose record of five victories he equalled, Belgian Eddy Merckx and France's Bernard Hinault, along with the late Jacques Anquetil the only other riders to have won the Tour five times. Indurain, the only rider to have won five Tours in succession, refused to be drawn out about how many more he may add.

This was Indurain's 11th Tour. He failed to finish the first two, but finished 97th in 1987. From 1988 to 1995 he has finished 47th, 17th, 10th, 1st, 1st, 1st, 1st and 1st.

There is still a serious lack of studies focused on such companies. What explains innovative success at the level of the whole organisation? To use the cycling analogy again, there is a need for studies which focus on cyclists who can be found among the front-runners in almost every race: the winners of the 'Tour de France', the 'Giro d'Italia', the 'Vuelta a España' and so on. In practice, there is a clear need for studies which look at the organisational context in which innovation processes are being conducted and which describe generic instruments and tools to enable managers to keep their organisation successfully innovative for an extended period of time.

NOTE

1. Although metaphors help us to see and understand organisations only partially, they are very useful in visualising and highlighting certain aspects of a problem: the use of metaphors implies a way of thinking and seeing that pervades the manner in which we understand our world in general (Morgan, 1986, p. 12).

2. Theoretical Framework

Nothing lasts forever, except change. (Heraclitus, 500 BC)

2.1 RESEARCH ON INNOVATION SUCCESS FACTORS

Over the past few decades we have witnessed a tremendous growth in studies related to 'innovation', and the attention is still increasing. A quick scan through 400 major economics journals indexed in the EconLit[1] database, for instance, showed that the number of innovation-related articles rose from 1026 in the 1970s (1970–79) to 4917 in the 1980s to 6307 in the first seven years of the 1990s (1990–96). Managing innovation has become a topic of interest for researchers from a wide variety of backgrounds. The topic is being studied from various angles and studies differ with respect to the sector studied, the level of aggregation (individuals, projects or firm), the size or type of company (high-tech start-ups, large conglomerates and so on), the scope (incremental, radical and so on) or type of innovations studied (product innovations, organisational innovations, process innovations) and the geographical setting (Grønhaug and Kaufmann, 1988). Its popularity and wide applicability resulted in a proliferation of meanings of the word 'innovation'. In general, the various meanings that the term 'innovation' has acquired over the years can be clustered into three main concepts (Zaltman et al., 1973):

- *The process of developing the new item.* This concept refers to the innovation process starting with research or market demand and developing towards widespread utilisation.
- *The process of adopting the new item.* This refers to the diffusion process, i.e. the process of user acceptance and implementation.
- *The new item itself.* This refers to the outcome of the development process, the new or improved product, service, process, management technique and so on.

When looking back on the research done in the past, we might state that the dominant focus in innovation research has gradually evolved *from the new item itself* and *the process of adopting the new item* towards *the process of developing it.* In the 1960s and 1970s, studies on the 'innovation' phenomenon concentrated primarily on the diffusion of innovation (Rogers, 1983) or on the technical aspects of innovation as a technological problem (for example Ayres, 1969; Allen, 1978; Beattie and Reader, 1971; Ford and Ryan, 1981; Kantrow, 1980; Stobauch, 1988; cf. Roussel, et al., 1991; and Saren, 1984). It was not until the mid-1980s (with the notable exception of the study by Burns and Stalker in 1961, which was clearly ahead of its time) that emphasis was increasingly put on *the organisational system* in which innovation processes take place. The characteristics of the innovative firm were studied (Moss Kanter, 1983) and organisational research identified managerial and organisational factors which inhibited or enhanced the success of innovations. Among other things, it was shown that centralisation is conducive to the adoption of radical innovations (Hage, 1980), and that firms with higher levels of top management team diversity display higher levels of innovation (Bantel and Jackson, 1989). Buijs (1984) demonstrated that a consultative (process) approach to the management of innovation projects is far superior to procedural or content-expert approaches. Furthermore, attention was paid to entrepreneurship (Ven, 1986) and intrapreneurship (Burgelman, 1983).

But an elusive phenomenon like innovation has of course not been the exclusive domain of organisational scholars. Innovation has become a subject of study in a wide variety of disciplines ranging from psychology, sociology and social anthropology to economics, public policy and marketing (cf. Rogers, 1983). There is, however, no unified 'theory of innovation'. And despite the fact that multi- or interdisciplinary co-operation is preached by the vast majority of researchers involved in innovation studies, innovation research itself is far from being truly interdisciplinary. Moreover, as Grønhaug and Kaufmann (1988, p. 4) point out, 'researchers preoccupied with innovation are partly unaware of the research done and conceptualisations used by colleagues from other disciplines'. Main concepts differ somewhat between disciplines and the same concept may be defined differently by various disciplines.

One of the most common definitions of an innovation is 'renewal with respect to products, markets and (technological) production processes' (Buijs, 1984). By defining innovation as discontinuous changes in the product/market/technology combinations of existing firms, Buijs (1984, p. 31) emphasises the discontinuous character of innovation. The question then is how 'new' the renewal must be or how 'discontinuous the change' for it to be considered as an innovation. The answers to these questions depend on the object of study. Studies on innovation diffusion, for instance, tend to accept the label 'new to the world' as the differentiating criterion. From an economist's point of view this might perhaps be the best criterion on to study. However, innovation in the vast majority of firms stops well short of 'new to the world' or even 'new to the industrial sector'. Most innovations in the average firm (and especially the SME) are variations on known themes. This often involves translating an idea or product from one field of application or market to another.

Thus, when innovation management within the individual firm is the object of study, defining an innovation as 'something new to the world' is not appropriate. In fact, companies themselves tend to use a more relaxed definition. In various studies preceding the current one, we interviewed numerous managers about the innovative efforts of their companies. When asked how 'new' a new product or service developed internally had to be in order to be considered an innovation in their firm, most replied 'new to our firm'. Only very few managers regarded 'new to the world' or 'new to the manufacturing or service sector' as their criterion. And this is quite understandable from the perspective that changes within the company (even smaller ones) bring with them uncertainties and dilemmas that have to be managed. And since the focus of our research is on managing the processes by which these new or improved products, services or processes are developed, 'new to the company' can be regarded as the most appropriate criterion.

2.2   LEVELS OF STUDY

Innovation, or the renewal of products, processes and services, can be studied at various levels. Three levels[2] are of particular importance in a study which aims to detect the reasons for differences in innovative success among firms, namely:

1. the firm's environment (the external organisation);
2. the firm; and
3. the project.

Each level will be discussed in greater detail in the following sections.

2.2.1   The environment or external organisation

External organisation theory regarding innovation has developed according to two broad traditions. One tradition sees innovation as an object of study for industrial organisation (a sectoral approach), while the other proposes a more geographical and infrastructural (regional) approach.

Sectoral approaches

Various innovation studies treat innovation as an industrial-economic phenomenon (cf. Scherer, 1982, 1975; Mansfield, 1989). In those studies, the emphasis is on market conditions, governance, R&D cartels, joint ventures and other relations between firms of a certain manufacturing or service sector. For instance, Scherer (1982) discovered strong interweaving in R&D activities between certain branches of industry. Other sectoral explanations for innovation include the extent of concentration, the capital intensity of the sector and the degree of vertical integration within the value-added chain.

Regional approaches

An infrastructural or regional analysis is also aimed at a level beyond the individual company, but is focused on regional characteristics and differences rather than sectoral ones. Bolton (1971) attributed entrepreneurship, industrial renewal and innovations to the quality of the environment, determined by the physical infrastructure, the educational system, potential investors, a favourable fiscal climate, and so on. The Enright study (1989) focused on industrial districts where a complete branch of industry is concentrated in a small geographical area. Geographically clustered and complementary companies who have created an adequate infrastructure appear to be capable, despite their small scale, of building and maintaining a strong global competitive position. The growth pole theory of Perroux is orientated to the positive effects that large firms can bring about in a region (Hartgers et al., 1990; Corvers et al., 1994). The underlying reasoning is that the establishment of a technologically advanced and rapidly growing large innovative enterprise can become a dynamo for regional development; especially if this firm has many input–output relations with other firms.

Apart from these vertical networks, a region can also blossom through strong horizontal networking, as shown for instance by studies on local systems of innovation (Lundvall, 1992) and on innovation network learning dynamics (Swann, 1993). The success of regional economic development in regions like Emilia Romagna (Italy: clothing industry) and Baden-Württemberg (Germany: light and mechanical engineering industry) can be explained in particular by the creation of bottom-up developed networks between medium-sized and small firms and between firms and organisations in the technological infrastructure such as research centres, educational facilities and so on (Corvers, et al., 1994).

Policy makers are increasingly appreciative of the value of bottom-up and need-driven innovation infrastructures and strategies. The current EC regional innovation support schemes (RITTS, RIS and the former RTP[3]) are examples of public schemes focusing on regional innovation strategies which are aimed at supporting local and regional governments in the (re)design of the regional innovation, technology transfer and R&D infrastructures; and which are targeted to ensure a match between the innovation-related needs of the SMEs and the services provided by the regional infrastructure. They are initiated and co-financed by Directorate-General (DG)XII and DGXVI and executed by regional authorities in their respective regions. Generally, these assignments involve a thorough assessment of the technology requirements and local needs, capabilities and potential, including management, financial, commercial, training and organisational issues as well as purely technological ones. The goal is to develop more efficient innovation support and promotion policies in the regions concerned. The resulting strategies generally provide a framework for optimising innovation policy and infrastructures at the regional level, especially with regard to their relevance to the needs of small and medium-sized enterprises.

2.2.2   Research at firm level

A shortcoming of the industrial-economic research discussed in the previous section is that it regards the innovating organisation as a black box. Internal processes within the innovating organisation are studied only briefly or not at all. As a result, internal constraints and innovation processes are mostly neglected. These issues are more the domain of studies focused on the firm level in which 'the innovating organisation' itself is the focus of study. *The starting point is that differences in the structure, culture, control and management of innovative activities lead to differences in innovative success.* Organisations provide the context within which innovations sprout, blossom or become adopted. The adoption of new ideas represents a major challenge for organisations since any innovation represents a departure from their current practice. Several theories on how organisations can be conducive to successful innovation have been advanced. The study by Burns and Stalker (1961) dealing with change processes has become a classic in innovation research. It provides one of the first theoretical statements on organisational innovativeness. They identified organisational attributes that distinguished effective innovators from organisations that could not match their performance. Burns and Stalker state that mechanistic organisations have more difficulty in implementing innovations than do organic ones, and that mechanistic organisations in particular, face poor survival prospects when market conditions no longer sustain old products and services. Their work essentially posits a contingency model in which successful innovators require a decentralised structure with informal working arrangements. In Chapter 7, we shall elaborate more on their ideas.

There is a widespread conception that innovative success can be attributed to the core activities and resources of the firm (cf. Penrose, 1959; Moss, 1981; and Wernerfelt, 1984). In this view, while the knowledge and skills embodied in an organisation can be a lever to renewal, they also define the borders within which the firm can successfully expand its core activities. Nelson and Winter (1982), for example, state that a company's functioning is to a significant extent determined by its routines. They compare routines with the genes which determine the behaviour of living beings. However, unlike genes, they argue, routines can be changed or replaced.

The importance of collaborative relationships between the key organisational functions involved in innovation processes has been the topic of many studies (Cooper, 1986). Since each innovation requires the specific and, to a certain extent, unique creation of such linkages, it is not surprising that organisations display both success and failure over time. According to Dougherty (1992), for instance, innovations can fail because certain conditions have not been met. These conditions include horizontal linkages between functional departments and between the firm and its customers and the presence of vertical linkages between strategic decisionmakers and those who implement the innovations. Earlier studies that have examined innovative organisations rather than innovation projects include Zaltman et al., (1973), Aiken and Hage (1971), Mohr (1982) and Moss Kanter (1989).

2.2.3 Research at project level

Much of the previous research on innovation success factors has, however, been focused on projects within organisations rather than on organisations as a whole. The SAPPHO studies (Rothwell et al., 1974; Freeman, 1982), some of the first success-versus-failure studies, identified a pattern of differences between a paired sample of 43 successful and unsuccessful projects. It was shown that the successful projects differed from the less successful ones with respect to the understanding of user needs, the attention paid to marketing and publicity, the efficiency of development activities, the use of outside technology and scientific advice and certain characteristics (such as age and authority) of the responsible individuals. The projects did not differ with respect to the pace of innovation. The initial SAPPHO study focused on two industrial markets in Britain: chemicals and scientific instruments. Similar studies have been conducted in the Hungarian electronics industry (Rothwell, 1974) and the textile machinery industry (Rothwell, 1976). They all yielded comparable results.

Rubenstein's (1976) study on factors influencing the success of innovation projects attributed much importance to facilitators of innovation such as product champions, effective internal communications facilities and superior techniques for data gathering, analysis and decision making. The NewProd project (Cooper, 1979; 1980) identified key factors distinguishing between successful new products and failures (in terms of commercial success). Success factors were: product superiority; a strong market orientation; and technological suitability and proficiency. Most of the success factors he identified are controllable by management. He therefore concludes that success is not nearly as dependent on uncontrollable or situational factors (such as the market, the nature of the company or the type of project) as might have been expected. In another study, Cooper (1983) identified seven distinct types of innovation processes and showed that new product success depends to a large extent on the way a firm conceives, develops and commercialises a new product: the new product development process. The profile with the highest success rate (in terms of success versus failure and overall programme performance) featured a balance between marketing-orientated and technically orientated activities and was characterised by a large number of varied steps in the process.

The Stanford Innovation Project (Maidique and Zirger, 1984) underlined the findings of the SAPPHO and NewProd studies and attributed new product success to in-depth understanding of the customer and the marketplace, the attractiveness of the marketplace, the marketing capabilities of the firm, the uniqueness of the product, the planning and co-ordination of the development process and the selection of products with markets and technologies that benefit significantly from the firm's present strengths.

2.3 THEORETICAL FLAWS

When looking at the vast amount of studies on innovation management, we can detect a few broad research discrepancies.

2.3.1 Project-orientated studies

As indicated in the previous section, research on managing innovations has predominantly been focused on the project level. The project level has attracted overwhelming attention from researchers into innovative success; and rightly, for various studies (Ven, 1986; Quinn, 1985) have shown that the success of an innovative firm is to a significant extent determined by the manner in which the innovation processes are managed. A study of successful and less successful innovative companies will thus have to devote some attention to the management of innovation projects as well. The question, however, is whether too little attention has been devoted to the whole organisation as an innovative entity. The scope of the research at organisational level is in glaring contrast to the considerable quantity of publications focused on the success factors of individual projects. Studies focused on the achievement of innovative success at company level have been quite scarce. Yet in order to establish a conclusive link between company attributes and effective innovation, we need to move beyond the analysis of a single innovation project. The reason for this is that innovative success at company level does not exclude innovative failure at project level. It is quite likely that successfully innovating firms might have had some projects which turned into absolute failures. It might also be the case that successfully innovating firms owe a large part of their success to a risk-taking project-selection strategy. One can imagine that some firms allow a certain number of high-risk, vaguely defined projects to emerge: projects which sin against all the 'success criteria' found in the project-level studies and which might have a low success rate. But when one of them becomes a success it could be a real triumph. Hence a high-risk aversion at project level might be complemented by a low-risk aversion or experimental tendency at firm level. Studies at project level give no insights into these matters. Knowing what makes projects succeed, therefore, can only partially answer the question as to what makes innovative firms succeed. And vice versa. The question as to how a firm can become a successfully innovating firm can therefore not be answered by just looking at the project level. Maidique and Hayes (1984) argue that a project cannot be seen separately from other projects over time and state that the literature indicates that project success is embedded in the rest of the company. Knowledge about innovative success at company level can thus lead to more knowledge about achieving innovative success at project level, and vice versa.

2.3.2 Manufacturing-dominated studies

In recent decades, many economies have increasingly evolved towards service-intensive economies. By now, the overwhelming majority of workers in many West European countries are employed in the service sector. In the Netherlands, the service sector accounts for over 67 per cent of employment (Bilderbeek and Hertog, 1992). Statistics show that the service sector accounts for 60 per cent of the value added in the European Community (Ghobadian et al., 1994). And the importance of the service sector is still rising. This increased importance of the service sector is, however, not reflected in the attention devoted to it in studies on successful innovations. These have shown an undue preoccupation with manufacturing firms (Booz, Allen and Hamilton, 1982; Maidique and Zirger, 1984, 1985; Johne and Snelson, 1988; Link, 1987; Cooper, 1990; Cooper and Kleinschmidt, 1993a).

The notion of (technical) innovation as an important resource for economic development seems to be predominantly established in manufacturing industries. According to this conception, the service sector is a derivative function that plays no independent innovative role of significance (Bilderbeek and Hertog, 1992). Yet innovation is not only discernible within the manufacturing industries or high-tech sectors. Service-providing firms face as much of a challenge in maintaining their capacity to successfully implement innovations as do manufacturing firms. And incorporating new technology is as important an issue for service-providing firms as it is for manufacturing firms. Some service sectors (such as banking, telecommunications and transportation) are among the most prolific users of information technology.

2.3.3 High-tech dominance

Moreover, the overwhelming emphasis of innovation studies in manufacturing industries has been on the high-tech firms, perhaps because issues related to new product creation or new process technology are most salient there. Consequently, much research has focused on microelectronics and, more recently, biotechnology. A similar phenomenon can be observed in the service sector. Many innovation studies in the service sector have focused on sectors characterised by intensive high technology, such as the financial services (Easingwood and Storey, 1991; Cooper and de Brentani, 1991; Cooper et al., 1994). But innovation is certainly not the exclusive concern of high-tech firms. Innovation is not synonymous with technology, but synonymous with change. Sometimes, the most profitable changes can even be achieved with existing technologies. For instance application innovations (which are hardly based on new technologies but which offer a substantial improvement in the perceived benefits to customers) are among the most successful kinds of innovation (Gobeli and Brown, 1987). Radical and technical innovations, on the other hand, show much lower overall success rates.

2.3.4 Size-related studies (limited to specific size types)

When considering the decisive factors for innovative success which can be influenced by the company, one must inevitably make a distinction between the small and medium-sized enterprises (SMEs) and the large multidivisional companies. There are simply a number of specific problems which have to do with the size of the firm. In large companies, for example, the isolation of the R&D function frequently poses problems, while innovation in small firms is usually pushed aside by the hectic day-to-day routine. In consequence, research on innovation success factors has focused predominantly on either the large conglomerates (with the likes of 3M, Intel and IBM as exemplars of firms that succeed or fail in their efforts to pursue innovations) or the small high-tech start-ups (the Silicon Valley type of firm). Small and medium-sized enterprises (in European terms, between 20 and 500 employees) are hardly covered at all. Yet most of the growth in employment and added value can be attributed to medium-sized and small firms in low- and medium-technology sectors. Several of these sectors are in manufacturing, but many others are in services. And innovation is as important to them as to the large firms. The demand for knowledge on innovation in SMEs is spurred on by the increasing attention in many regions to designing SME-need-driven regional innovation strategies (Cobbenhagen and Severijns, 1999). A development that (in Europe) is stimulated by the RIS and RITTS initiatives of the European Commission (see subsection 2.2.1).

2.3.5 Sector-specific studies

Many sectoral studies on innovation success and failure have been conducted in a single sector, such as the textile machinery industry (Rothwell, 1976), the electronics industry in Hungary (Rothwell, 1974), the electronics industry in the US (Maidique and Zirger, 1984), the pharmaceutical industry (Bower, 1993), and the German mechanical engineering industry (Murman, 1994). Other studies have focused on comparing two sectors, such as the chemicals and scientific instruments industries (Freeman, 1982), the food processing industry and the medical instruments industry (Karakaya and Kobu, 1993), and the Dutch agro and food industries (Sneep, 1994). Research in the service sectors shows a similar picture: many studies focused on just one or a few sectors. And this quite frequently included the financial industry: for instance, new product success in the UK financial market (Easingwood and Storey, 1991) and in the financial service industry (Cooper et al., 1994). Research on non-sector-specific innovation success factors has, however, been very limited. A notable exception is Selznick, one of the pioneers of the core competency school. Selznick (1949, 1952, 1957) studied organisations as widely divergent as the Tennessee Valley Authority and the Communist Party with the aim of finding distinctive characteristics (distinctive competencies) which account for success.

And of course, there are good arguments in favour of sector-specific studies since *some* innovation success factors are idiosyncratic to the specific environment of the firm. These include, among others, concentration ratio, entry barriers and exit barriers. And these environmental characteristics have their influence on the characteristics of innovation processes within the firms.

Pavitt (1984), for instance, describes and explains sectoral patterns of technical change. He then classifies the characteristics and variations in a three-part taxonomy based on firms: (1) supplier-dominated; (2) production-intensive (with a distinction between scale-intensive firms and specialised suppliers); and (3) science-based.

Sectors differ with respect to average throughput times for innovation processes. The development of a pharmaceutical drug, for instance, takes considerably longer than an innovation in the furniture industry. And legislation places very specific demands on firms which are reflected in the design of innovation processes. Managing the development of a new drug as if it were the development of a new couch would therefore lead to disaster. In the supplier sector, the *direct relationship* (co-operation) with the customer is of great importance, as is market knowledge and research in the consumer sector. In the pharmaceutical industry, basic research and clinical trials play a central role. Production and sales are sometimes even regarded as a means of providing money for research. In a machine plant, process renewal, craftsmanship and the integration of process technology and the organisation are the central elements. It can therefore not be denied that the design of the innovation process is partly determined by the nature of the market, the product and the process. Sometimes, the differences within one sector are so large that firms have trouble serving both markets. An example of this has been observed in the chemical industry, where speciality chemicals and commodities require a completely different innovation focus (Cobbenhagen and Philips, 1994).

At first glance, then, there is much to say for the argument that innovative success depends to a large extent on sector- or even company-specific factors. In a sector study, however, it is difficult to see which success factors are sector-specific and which are general. These studies only show us that the factors found seem to work in that particular sector. One intriguing question thus remains unanswered: *Do the same concepts regarding good innovation management apply to an insurance firm as well as to a chemical firm?* The question as to which success factors are idiosyncratic and which are general can only be answered by studying the phenomenon from both angles. There is, however, a surprising lack of studies which allow for the transfer of best-practice lessons from one sector to another.

Furthermore, we must note that many arguments in favour of sector-specific studies on innovation management are related to the study of the management and control of successful innovation *projects*, not successfully innovating firms. But when one considers the latter, more general, level of the innovating firm, it is very likely that a large number of non-sector-specific success factors could be identified. In fact, Rumelt (1991) found that business units differ far more within than between industries with respect to factors which explain variations in success. *We therefore hypothesise that many success factors are less industry- or sector-specific than is often asserted.*

2.4   BASIC RESEARCH QUESTION

This book is based on the research project entitled 'Successful innovating firms: breaking away from the pack', initiated by the Dutch Ministry of Economic Affairs.[4] The question posed by the Ministry of Economic Affairs was the following:

> Some firms have gained a lead on their competitors in terms of continuous innovative success. In what way do these firms differ from their competitors in the primary activity which can be referred to as innovation management? And which lessons can be learned from this comparison with respect to the business policies of companies which are less successful in their innovative attempts (members of the pack)? Lessons which apply to a wide range of manufacturing and service sectors.

As discussed in previous sections, there is a vast amount of literature on innovation and innovative success which has focused on one or a few sectors. The current study should therefore focus on the question as to whether the concepts which originated from these studies or other concepts could be applied in a broad range of industries and service sectors and result in the same conclusions. The main research goal, therefore, is to provide non-sector-specific (or industry-neutral) insights which can improve the innovative success of firms. Preferably, these factors should be controllable by the firm. We have to keep in mind, though, that the resulting success factors might lead to general guidelines which need further refinement per sector or even per firms; but this is inherent in any form of meso- or macro-level research.

Since this study focuses on innovative success at firm level, I have defined innovative success as the ability of organisations to generate a continual stream of positive incomes from its innovative efforts. This goes beyond the notion of innovative success as a simple tally of innovations, which might be the conclusion if the research were focused at project level. If we view innovation as a cycle race (Figure 2.1), the research question can be formulated as follows:

Basic research question

Which non-sector-specific factors determine the difference between the front-runners (successfully innovating firms) and members of the pack (firms with an average success rate for their industry in their innovative efforts) with regard to innovation and how can these factors be weighed?

Laggards   The pack   Front-runners

Figure 2.1 Innovation as a cycle race

The present study is founded on two pillars, a practical and a scientific. The practical pillar is intended to provide firms with clear insights and tools to improve or sustain their innovative success. This part of the study has led to various practice-orientated publications (Cobbenhagen et al., 1994a; 1994b; 1995; Cobbenhagen and Hertog, 1995; and Hertog and Cobbenhagen, 1995).

The current book is the account of the scientific research stream within the study. This part of the study is aimed at contributing, in a modest way, to the development of a pragmatic framework that provides insight into non-sector-specific success factors of innovating firms.

It is important to note that the study is not focused on the differences between the best-performing firms (the front-runners) and the worst-performing (the laggards), but on the differences between the best and the industry-average firms (the members of the pack). Comparing the best to the worst would obviously lead to much larger (and more significant) differences, but it is quite conceivable that this comparison might reveal more about good management in general than about good innovation management. Furthermore, there is a practical reason for selecting the group of pack members as the contrasting group to the front-runners; and that is to provide the pack members with suggestions as to how to escape from their position and become front-runners. And front-runners are provided with insights which allow them to keep ahead of the pack. Although it is very doubtful that laggards can become front-runners, it is quite conceivable that pack members can break out and join the front-runners.

2.5   THEORETICAL FRAMEWORK

Much of the previous literature on critical success factors is aimed at identifying factors (characteristics of projects, companies, people and so on) which may explain differences in success. However, most often it will be a combination of factors rather than individual factors that account for (differences in) success. Here, I shall describe these combinations of factors in terms of competencies. This will be done by building a link between strategy literature and innovation literature. To that extend I first discuss two paradigms from the strategy literature.

2.5.1   Paradigms in strategy analysis

In this subsection, two dominant paradigms in strategy analysis are discussed: the 'strategy fit and allocation' paradigm and the 'strategy as stretch and leverage' paradigm. Although they are sometimes regarded as opposed to each other, it is argued that they in fact supplement each other and should be regarded as complementary.

Strategy fit and allocation
The core of the traditional strategy school (Chandler, 1962; Ansoff, 1965) is formed by the link between strategy and the external environment: the so-called 'strategic fit'. According to this 'strategy fit and allocation' paradigm, companies should position themselves competitively on the basis of a fit between the company's strengths and weaknesses on the one hand and the chances and opportunities in its environment on the other (Gorter, 1994). Until the 1980s, the principal developments in strategy analysis remained focused on this strategic fit and allocation paradigm (Grant, 1991, p. 114). Two prominent examples of this paradigm are Porter's positioning school (Porter, 1980) and the empirical studies undertaken by the PIMS project (Buzzel and Gale, 1987).

Porter (1980) argues that, based on a thorough analysis and understanding of the industry structure, a company has to choose one of the following three strategic positions: cost leadership; differentiation; and focus. This 'outside-in' approach has dominated the strategic agendas of consultants, researchers and managers for over a decade (Wit, 1994).

However, the positioning school did not prove to be the 'recipe for success' that was hoped for. Awareness of the limitations of one-sided concentration on positioning and excessive external focus grew. After all, the 'strategy fit and allocation' paradigm focuses attention primarily on what other people do. Subsequently, one's own strategy is determined on this basis. It establishes what an organisation *'should want'* to do. The question as to whether the organisation is capable of doing so remains unanswered. And a company's capacity to distinguish itself lies precisely in *that capability*. In the recent literature, a clear reversal can thus be observed from the analytical, relatively static concept of 'strategy fit and allocation' to the more dynamic, business-focused concept of 'strategy as stretch and leverage' put forward by scholars such as Hamel and Prahalad (1989).

Strategy as stretch and leverage

Hamel and Prahalad (1989) doubt whether it is possible for firms to analyse the success of other companies and apply this knowledge for their own purposes. Companies which base their policies on market and competition analyses will ultimately differ little from each other and thus will become involved in fierce competitive struggles. By excessively gearing their own strategy to that of their competitors (the market leaders in particular), companies run the risk of falling behind time and time again (Hamel and Prahalad, 1989). The authors introduce the term 'strategic intent' to stress their proposition that company success is based on high ambitions. They argue, for instance, that companies which have risen to global leadership over the past two decades invariably began with ambitions that were out of all proportion to their resources and capabilities. These companies created an obsession with winning throughout all levels of the organisation and were able to sustain that obsession during the long quest for global leadership. In their view, company success cannot be the result of strategic planning, since this implies the existence of rules, lines of sight or patterns, which are only valid when all 'players play the same game and stick to these rules'. Hamel and Prahalad use the success of Canon to illustrate the threat which can be posed by a newcomer which does not live up to the existing 'rules'. Canon's 'obsession' was to 'beat Xerox'.

In the past decade, we have therefore witnessed an increase in interest among scholars and practitioners in the role of the firm's resources as the foundation of its strategy and success. Hamel and Prahalad (1989) argue that the importance of market share as a traditional measure of market power and leadership will lose ground to a concept called 'manufacturing share', that is, the ability to control core competencies (Prahalad, 1990). In general, this strategy as stretch and leverage paradigm is founded on pioneering long-term company goals based on a clear vision and supported by strong ambition throughout the organisation. Long-term goals force companies to take major steps (stretch) and deploy their resources and capabilities with the utmost effectiveness and efficiency (leverage). In this process, firms are not guided by the available space within the branch of industry, but (Gorter, 1994) generally create new product-market combinations. An important basic principle in this line of reasoning is the proposition that the competitive advantage of firms stems from *dynamic capabilities rooted in high performance routines operating inside the firm, embedded in the firm's processes, and conditioned by its history* (Teece and Pisano, 1994). Teece and Pisano are co-founders of the resource-based school, a prominent exponent of the strategy as stretch and leverage paradigm. Thus, according to this paradigm, the success of a company is dependent on its ability to activate its own resources. Company (innovation) strategies must therefore be grafted onto these crown jewels. This way of thinking is diametrically opposed to the classical strategic thinking of the strategy fit and allocation paradigm, which is primarily focused on the external environment.

Strategy fit and allocation versus strategy as stretch and leverage
The two paradigms have a number of fundamentally different basic principles. The three most important ones are related to strategic freedom of choice, possibilities for co-operation and customer orientation.[5]

Strategic freedom of choice A significant difference between the two paradigms has to do with their basic principles regarding the strategic freedom of choice available to companies (Wit, 1994). The strategy fit and allocation paradigm is based on a relatively deterministic standpoint. The strategic freedom of choice of firms is limited and strategy is to be primarily focused on keeping the competition at bay. The firm's environment is considered to be more or less a given factor. From the perspective of the stretch and leverage paradigm, on the other hand, companies have a great degree of freedom to determine their strategies. The competitive environment, according to this point of view, is not a constant factor. On the contrary, it is a result of strategic activities conducted by firms and is thus subject to influence. According to this school, an external position is not something which is selected following rational analysis of the environment. Rather, it is created on the basis of unique capabilities and core competencies, both those which are already present and those which are yet to be developed. It is an inside-out approach, while the strategy fit and allocation paradigm is an outside-in approach.

Possibilities for co-operation A second important difference in assumptions concerns the firm's possibilities for co-operation (Wit, 1994). A significant implicit assumption in the strategy fit and allocation paradigm is that the competitive environment consists primarily of enemies. From this point of view, the environment is competitive and a clear dividing line can be drawn between the firm and the environment. Co-operation with other companies is not very popular and is considered to be a sign of weakness, a 'last resort'. In contrast, the strategy as stretch and leverage paradigm sees co-operation as an effective manner of creating new opportunities. Through co-operation, companies can acquire or develop knowledge, penetrate new markets, learn from other companies or make a joint stand against a common threat. The firm's boundaries are much more virtual from this perspective.

Customer orientation The core competency idea is revitalising the entrepreneurial spirit. Strategy is again becoming aggressive. The servile attitude of customer orientation, patiently listening to wishes and needs or filling small gaps in the market as opportunity arises, does not fit within this context (Witteveen, 1994, p. 123). A different approach to the environment and the customer is being taken. In the process, the position of the business unit structure focused on product-market combinations is being called into question. Prahalad and Hamel see business unit formation primarily as a morbid growth which has developed out of the traditional strategic approach. In their view, rather than reinforcing market orientation, the division of the organisation into business units will actually weaken the company, since resources for the development of core competencies will also be fragmented in consequence. Along with Gorter (1994, p. 129), we feel that this goes a bit far. According to Hertog et al. (1996), an organisation must learn to deal, in terms of structure, with the paradox inherent in the fact that a firm must facilitate its external orientation on the one hand (market and customer orientation) and its internal strengths (development of core competencies) on the other. The first orientation benefits more from a business unit structure, the second from a concentration of knowledge in an organisational entity. In current practice, we can identify innumerable examples of such a hybrid structure. DSM, for example, has for some time now had a corporate R&D programme and a technology council which, at concern level, monitors the technological knowledge which goes beyond the bounds of the business units. The council is also responsible for effecting the transfer of knowledge between business units. In addition, there is a business development unit which functions as an incubator for new developments which do not fit so well within one of the business units.

Although the strategy as stretch and leverage paradigm and the strategy fit and allocation paradigm differ on a few important basic principles, we believe that they should not be regarded as mutually exclusive. The two, in fact, are complementary on many points, 'with one explaining the value of competitive outcomes in the product market, the other the dynamic aspects of firm behaviour with regard to the accumulation and disposition of the firm's resources' (Collis, 1991, p. 65). Together, they represent a combination of internal and external analysis which has been identified as the basis of good strategy formulation (Andrews, 1971). A recent study in the EC on the use of innovation management tools and techniques in stimulating innovation in SMEs has shown once more the importance of both external (outside-in) and internal (inside-out) assessments as a drive for change (Brown and Cobbenhagen, 1997).

2.5.2 The resource-based school

In line with the resource-based theory, we base our study on the premise that the success of an innovating firm is to a large extent determined by its resources and capabilities. The current study will therefore build on the concepts derived from the strategy as stretch and leverage paradigm in general and the resource-based school in particular. It intends, in a modest way, to contribute to the base of empirical work on capabilities and competencies, which, despite renewed theoretical interest in these ideas from the resource-based school, is still at a preliminary stage, as Henderson and Cockburn (1994) argue.

Origin of the resource-based school
The dissatisfaction with the static, equilibrium framework of industrial organisation economics that has dominated much contemporary thinking about business strategy has sprouted a revival of interest in older theories of profit and competition (Grant, 1991). While the notion that competitive advantage requires both the exploitation of existing internal and external firm-specific capabilities and the continuous development, exploitation and protection of firm-specific assets has not been a dominant view in industrial organisation (Teece and Pisano, 1994, p. 553), it nevertheless has a long tradition going back at least to Schumpeter (1934). Schumpeter regarded economic development as a process in which entrepreneurs dipped into a stream of technical opportunities apparently created for reasons independent of particular markets and brought those innovations to the market.

In describing the character of an organisation, Selznick (1957) used the term 'distinctive competency' to refer to those things that an organisation does especially well in comparison to its competitors. Selznick (1949, 1952, 1957) studied organisations as widely divergent as the Tennessee Valley Authority and the Communist Party, and in each case he noted the emergence of special capabilities and limitations as institutionalisation proceeds (Snow and Hrebiniak, 1980). Andrews (1971) further developed the concept of distinctive competency by showing that it was more than what the organisation could do; it was the set of things that the organisation did *particularly well* relative to its competitors. Distinctive competency is therefore an aggregate of numerous specific activities that the organisation tends to perform better than other organisations within a similar environment. Penrose (1959, p. 75) argues that 'It is the heterogeneity ... of the productive services available or potentially available from its resources that gives each firm its unique character'. A firm may achieve profits not because it has better resources, but rather because the firm's distinctive competency involves making better use of its resources (ibid. p. 54). Inspired by these ideas and further work by scholars such as Selznick and Penrose, recent publications have suggested that the possession of unique 'competencies' or 'capabilities' may be an important source of consistent strategic advantage. Current managerial literature by Irvin and Michaels (1989), Wernerfelt (1989) and Prahalad and Hamel (1990) further develops the notion of core skills and capabilities as a source of competitive advantage.

The resource-based school (also referred to as the dynamic capabilities approach) which has been further developed in Teece (1982), Wernerfelt (1984) and Teece and Pisano (1994), follows in Schumpeter's footsteps. However, it emphasises organisational processes inside the firm more than Schumpeter did. Furthermore, it is more prescriptive for companies searching for ways to change, while Schumpeter's approach is more descriptive.

Resources and capabilities

The resource-based school proposes that strategy be based on internal strengths. Two types of strengths are identified: *resources* and *capabilities*. Resources are defined as *the inputs into the production process* (Grant, 1991, p. 118) and include items of capital equipment, skills of individual employees, patents, brand names, finance and so on. Resources therefore determine what a company can do to a large extent, but do not determine success as such. In fact, there is no predetermined direct functional relationship between the resources of a firm and its capabilities. Nevertheless, resources (in terms of types, amounts and qualities available to a firm) do place constraints on the range of capabilities that can be performed and the standards to which they are performed (Grant, 1991). If no budget is allocated to innovation-related resources and the proper equipment is not available, it is clear that it will be hard to be innovative. But just spending money and having the resources available is also not enough to be successfully innovative. It is the manner in which the resources are applied that determines success. The organisation's style, values, traditions and leadership represent critical forms of encouragement to achieve this co-operation. Thus, in order for resources to be productive, 'teams' of resources have to co-operate and be co-ordinated.

Resources are therefore the source of a company's capabilities, a capability being defined as 'the capacity for a team (collection) of resources to perform some task or activity' (ibid.). Capabilities are heterogeneous, intangible, imperfectly transferable, firm-specific, resource-based and closely related to distinctive and core competencies. Capabilities are regarded as the primary sources of the company's competitive advantage. The essence of strategy is to make effective use of these internal crown jewels. A company which has unique resources and capabilities at its disposal and knows how to combine them in a relatively inimitable way will acquire a competitive edge. When capabilities are further combined or developed in areas of specialised expertise which endow a firm with distinct strategic advantages, these combinations of capabilities are often referred to as core competencies. Prahalad and Hamel (1990) use the term 'core competencies' to describe these central, strategic capabilities which are crucial for the success of the company, defining them as 'the collective learning in the organization, especially as to how to co-ordinate diverse production skills and integrate multiple streams of technology'. Competencies will be discussed in section 2.5.4. But first, the notion of capabilities will be discussed in greater depth.

2.5.3 Capabilities

Creating capabilities is not simply a matter of assembling a team of resources. Rather, it involves complex patterns of co-ordination among people and between people and other resources. Perfecting such co-ordination requires learning. Capabilities are not static. They are dynamic entities that change over time. A change in the capabilities of a company implies a learning process. This learning process is not necessarily conscious or intentional (Huber, 1991, p. 89). New capabilities have to be learned and old ones have to be unlearned, and for many innovating organisations this is a continuous process. This type of organisational learning[6] is regarded as a key strategic variable that drives innovation (McKee, 1992; McGill et al., 1992). It allows companies to adapt themselves in the face of rapid technological change. Companies learn internally by accumulating a wide range of knowledge, but they also learn externally through the process of adapting to external conditions. Only firms which have *learned how to learn* can stay successful for a long time, for they have learned to adapt quickly while maintaining a certain direction and identity (Swieringa and Wierdsma, 1990). At this point, it is important to stress that not all learning is good. For instance, learning does not always increase the learner's effectiveness, or even potential effectiveness. Furthermore, entities can incorrectly learn, and they can correctly learn what is incorrect (Huber, 1991).

The strategy fit and allocation paradigm is fairly deterministic in the sense that it is based on the notion that a company should adapt its strategy to circumstances. The strategy as stretch and leverage paradigm, on the other hand, is less deterministic. In this paradigm, it is recognised that organisations can choose their environment (by changing their market focus, relocating production facilities and so on) or even change it (through lobbying with policy makers, advertising and so on). The fit between the environment and capabilities is less strict, since capabilities can be used to change the environment. Furthermore, the time scale is longer. Nevertheless, success based on a certain set of capabilities is time- and situation-dependent and is not a guarantee for continued success. Organisations have to be prepared for these kinds of changes. History is full of numerous once-successful, organisations whose success turned out to be the cause of their decline. They became blind to changes and rigidly held to their once-successful way of operating. Lessons learned from past experience led to learning traps when the environment changed (Lant and Mezias, 1990). How dangerous an exaggeratedly positive self-image can be became apparent in the discussions generated by the publication of the book by Peters and Waterman on (1982) corporate excellence. Many of the 'excellent enterprises' in this book no longer seem to be worthy of that title (Stacey, 1993).

Dougherty (1992) discusses the organisational routines of successfully innovating companies which in time can become potential barriers to innovative success. In a later study, she uses the concept of 'core *in*competencies' to describe 'a scaffolding of rigid, objectified rules of thumb' which can grow around the core competencies of a company 'like vines run amok, trapping them and in some cases choking them off' (Dougherty, 1993, p. 31). Leonard-Barton (1995, p. 30) argues that 'the perplexing paradox involved in managing core capabilities is that they are core rigidities. That is, a firm's strengths are also – simultaneously – its weaknesses'. Core capabilities and core rigidities are two sides of the same coin. And often the flip side of core capabilities is revealed by external events. Then the internal pressure to stick to the core capabilities is great. Within the stretch and leverage paradigm, it is therefore argued that the capabilities forming the core competencies should be focused on the environment of the future. This is a different view than regarding capabilities as necessary for survival in the present environment.

2.5.4 Competencies

As stated in subsection 2.5.2, capabilities can be further combined and developed in areas of specialised expertise (core competencies) which endow a firm with distinct strategic advantages. These core competencies are defined as the outcome of collective learning in the organisation, especially with respect to the co-ordination of skills and the integration of technologies, and the result of harmonising complex streams of technology and work activity through a unique combination of technology, skills, people and resources (Prahalad and Hamel, 1990).

Defining an organisation in terms of its unique resources, capabilities and competencies is not new, however. Core competencies have been the focus of many studies in the past, albeit under different names, for instance: 'resource deployments' (Hofer and Schendel, 1978); 'distinctive competencies' (Snow and Hrebiniak, 1980) and 'invisible assets' (Itami, 1987). Although the definitions differ slightly, they all refer to the acquisition of a competitive advantage by excelling in certain capabilities. What is new in the current thinking on core competencies is the dynamic undertone of the core competency philosophy. Whereas distinctive competencies were more passive and related to competencies built up in the past, the core competencies are defined as dynamic competencies intended for future development (Hamel and Prahalad, 1991). The resource-based school (and the strategy as stretch and leverage paradigm in general) believes that the company's strategic intention to obtain a competitive edge is of crucial importance. Core competencies play an important role in the actual achievement of this strategic intention. They are unique, company-bound features which provide access to a wide range of markets, make a significant contribution to the product benefits perceived by customers and, moreover, are relatively immune to imitation by competitors. Core competencies are not the capabilities to which the company owes its present success, but rather the capabilities to which the company will owe its future success. Core competencies must thus be defined by management in a forward-looking manner and cannot be looked for in retrospect (Gorter, 1994). They are, as it were, the company's potential for success in its endeavours. Core competencies are therefore not limited to the present, but provide potential access to a wide variety of markets. Effective application of these core competencies allows firms to offer new products, develop previously unseen market opportunities and foster growth (Prahalad, 1990). Firms should not think of their competencies as mechanisms for cloning the past, they should rather focus them on the future. In fact, it may even be possible that a company's current success with its services or products is not based on the core competencies defined at the present time.

In their definition and further elaboration of the core competencies concept, Prahalad and Hamel (1991) focus on technology-related competencies. Managerial, organisational and marketing competencies are hardly discussed. In section 2.1, however, we saw that the current literature on innovation management (by authors such as Cooper, Souder, Clark and Wheelright) focuses particularly on the organisational aspects of innovation. In recent policy initiatives at regional level, as well, we can see that organisational and marketing aspects are receiving considerably more attention, at the expense of purely technological matters, particularly with respect to policy focused on average small and medium-sized enterprises. There is a growing awareness (Cannell and Dankbaar, 1996; Cobbenhagen et al., 1996; Cobbenhagen and Severijns, 1999) that for such firms (roughly 80 per cent of all European companies) access to technological knowledge and technology transfer represents considerably less of a bottleneck than organisational and marketing knowledge.

A question unanswered by Prahalad and Hamel concerns *the manner in which* core competencies can be achieved. They talk about mobilising the organisation, but give no indications as to *how* and by what steps this can be achieved. Knowing what core competencies are is only partially helpful in the process of defining and building them. With this 'how' question, we find ourselves rapidly confronted with organisational issues, with the literature of innovation and organisation.

We therefore argue that (core) competencies are not just technology-based and can be found in many business processes. Snow and Hrebiniak (1980), for instance, have identified distinctive competencies in general management, financial management, product research and development in one of their case studies. Amit and Schoemaker (1993) mention the following examples of capabilities: highly reliable service, repeated process or product innovations, manufacturing flexibility, short product development cycles.

In this book, we should like to forge a link between, on the one hand, the strategy as stretch and leverage paradigm found in the strategic literature and, on the other, the recent trends in the literature of innovation management. With this in mind, we should like to introduce here the concept of 'managerial competencies'.

In order to give this concept a place among the wide diversity of (sometimes overlapping) concepts currently used in the literature on competencies, we shall now introduce the nomenclature used in this book. In doing so, we follow the general trend in organisational and strategic literature which defines competency in broader terms than the classical concept of competency in psychological literature, where it is generally related to individual knowledge and skills. In this broader view, a hierarchy of competencies can be identified (see Figure 2.2).[7]

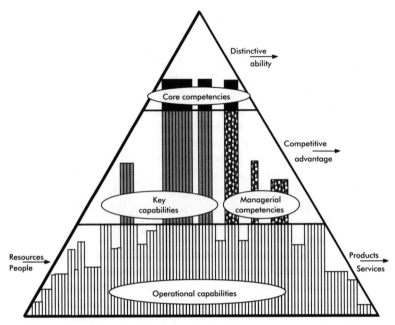

Figure 2.2   A schematic representation of the classification of capabilities and competencies

At the bottom of this schematic representation, resources and people 'enter' the system and products and services (the output) 'exit'. But products and services are not the only output of this transformation process. On a higher level of aggregation, the system can also lead to a competitive advantage or even a distinctive ability. Essential to this process is the creation of competencies and capabilities. These are built through a collective learning process within the organisation. Companies learn by amassing a wide range of knowledge, but also through the process of adapting to external conditions. This implies breaking old routines (habits and behavioural patterns) that are no longer appropriate, and learning new ones. The purpose of this change in attitude is to generate routines, capabilities and competencies which better serve the purpose of the organisation. Only firms which have *learned how to learn* can stay successful for a long time. They have learned to adapt quickly while maintaining a direction and identity (Swieringa and Wierdsma, 1990). It is therefore proposed that the system as depicted in Figure 2.2 be considered as a learning organisation in which inputs will increasingly more effectively lead to outputs. Through combinations of operational capabilities, the organisation will be able to develop key capacities and managerial competencies and thus obtain a competitive advantage (an output of the system at a higher level of aggregation). A further step in the learning process involves the development of core competencies with which a distinctive ability can be created. The four concepts used in this model will now be discussed in more detail.

Operational capabilities

At the bottom of the pyramid, we find the operational capabilities, which can be defined as the expertise and skills of people and groups within the organisation (such as research, marketing, distribution) which are applied, generally in combination with other capabilities, to achieve an output. Operational capabilities include both human capabilities as well as non-human resources. Although firms differ with respect to their operational capabilities, these capabilities are not distinctive. Operational capabilities provide no distinctive ability and companies can therefore hardly achieve a competitive advantage based on strengths in operational capabilities alone.

Key capabilities

Key capabilities can be defined as combinations of operational capabilities related to specific technologies, methods and techniques in which the firm excels or wants to excel. They include key technologies, key marketing techniques and so on, and provide the platform on which new products, processes or systems can be developed. Key capabilities are more firm-specific than operational capabilities. In fact, firms can achieve a competitive advantage by exploiting a unique combination of capabilities. In our view, key capabilities only provide an advantage when they are embedded in managerial competencies.

Managerial competencies

In addition to key capabilities (technological knowledge), managerial competencies represent a second important factor explaining competitive advantage. Intrinsic to the resource-based view of the firm is the belief that every firm must have the organisational capability to implement effectively its chosen product–market strategy and continually regenerate and upgrade its core competencies (Collis, 1991). The managerial competencies correspond to the distinctive competencies described by authors such as Snow and Hrebiniak (1980) and Amit and Schoemaker (1993). Snow and Hrebiniak (1980) cite the effectiveness of some organisations in managing product or market development as an example of distinctive competencies, that is, 'capabilities that their competitors do not have'. Managerial competencies can be regarded as an organisational counterpart of key capabilities. They are the management capabilities for effectively co-ordinating and redeploying internal and external competencies (Teece and

Pisano, 1994, p. 538), and they are less company-specific than the key capabilities. While key capabilities are focused on specific technologies or combinations of technologies, managerial competencies can be regarded as the glue which keeps it all together and the catalyst for achieving competitive advantage. The main goal of managerial competencies is to make sure that the key capabilities are tuned in to each other, so that the result becomes more than the sum of the parts. Managerial competencies go beyond simple management techniques and, like key capabilities, should be regarded as combinations of organisation-related techniques, attitudes, working methods and so on.

Core competencies

At the top of our triangle, we can identify the core competencies, defined as unique combinations of several key capabilities and managerial competencies applied in the firm's major business areas. A core competency has a wide application in the company's business, makes a significant contribution to perceived customer benefits, is difficult for a competitor to imitate and forms the basis for new business growth.

In this study, as we have said, we are making a cautious attempt to bridge the gap between the resource-based school and the literature of innovation management. For this reason, the object of study will be limited to the managerial competencies of companies. We are seeking to identify the managerial competencies which companies have built up in the past and which have enabled them to become successfully innovating companies. In addition, we are particularly interested in knowledge about managerial competencies and capabilities which is transferable to and useful for other companies. The underlying idea here is that success can be achieved by developing capabilities internally. The line of business and the environment are considered less significant factors. The question we shall be looking into is whether there is some kind of an organisational or managerial distinctive competence which explains why one company is more successful in its innovative attempts than another. And furthermore, whether there is a common managerial competence which can serve as a predictor for innovative success in a broad range of industries and service sectors. As will be seen in Chapter 4, a distinction will be made between technological, marketing and organisational managerial competencies.

As the study is aimed at practitioners as well as the academic community, my objective is twofold:

1. *The practical objective* In practical terms, I want to help meet the need for studies examining the organisational context in which innovation

processes are being conducted; and would also like to offer managers some general instruments and tools they can use to keep their organisations successfully innovative for extended periods of time. This will be done by providing knowledge about the critical success factors for innovative success.

2. *The scientific objective* Scientifically, I want to contribute to the development of new theory to enable us to explain differences in innovative success in a wide range of industries. To a large extent this will be done by constructing a model based on managerial competencies.

The first outlines of this model will be sketched in Chapter 4 on the basis of an intensive exploratory case study. The case study was needed to help work out the formulation of the study questions in both theoretical and practical terms. The model is intended to identify building blocks for managerial competencies and provide a tentative framework which links these building blocks and helps us to explain differences in innovative success. Chapter 3 will first deal with the methodological issues of the present study. Chapters 5–8 will discuss the study findings. Although a major part of the overall theoretical framework of this study has been discussed in the current chapter, more detailed discussions will be presented in these chapters. The overall discussion of the findings will be presented in Chapter 9.

NOTES

1. EconLit is a comprehensive, indexed bibliography with selected abstracts of the world's economic literature, produced by the American Economic Association. It includes coverage of over 400 major journals as well as articles in collective volumes (essays, proceedings, etc.), books, dissertations and working papers.

2. Apart from these three levels, one can distinguish the overarching societal level (which addresses questions related to technological change in general, patents, market structures and adoption by users) and the level of the individual (with questions related to such aspects as characteristics, skills and working styles of creative individuals, problem-finding and problem-solving techniques and techniques for stimulating creativity). However, both levels are beyond the scope of this book and will not be discussed. We refer to Grønhaug and Kaufmann (1988) for an overview of studies on these two levels.

3. RITTS = Regional Innovation and Technology Transfer Strategies and Infrastructure (DG XIII); RIS = Regional Innovation Strategies (DGXVI and DGXIII); and RTP = Regional Technology Plan Strategies (DGXVI and DGXIII).

4. Promotional support for the study was also given by two of the largest Dutch employers' organisations, AWV (Algemene Werkgevers-vereniging) and FME (Federatie voor Metaal en Electro).

5. Apart from these three important differences in basic principles, both paradigms differ with respect to various other dimensions as well. An extensive list of differences is provided by Liemt and Commandeur (1993, p. 114–115).
6. Various definitions of organisational learning are in use. Shrivastava and Grant (1985) define organisational learning as 'The autonomous capacity of organisations to create, share and use strategic information about themselves and their environment for decision-making'. Argyris and Schön (1978, p. 323) use a similar definition, but relate learning to improvement in performance: 'experience-based improvement in organisational task performance'. Hertog and Diepen (1989), link organisational learning to future success, defining it as 'the process by which an organisation makes use of information from past events to master future events better'. The common denominator of these definitions is that they emphasise the organisational interaction with the environment: changes in organisational modelling of the environment and the consequential organisational action (McKee, 1992).
7. The development of these ideas about core competencies and the hierarchical structure, is based upon the ideas developed at Philips as they were discussed during an interview with Prof. dr F. Meijer (Head of Corporate Research, Philips). Furthermore, proper credit should be given to Gallon and Stillman (PA Consulting) who wrote an internal document for Philips ('Putting core competency thinking into practice'). Their analysis of the core competencies phenomenon has triggered my own ideas.

3. Methodology

A path is created by walking on it. (Chinese saying)

3.1 RESEARCH STRATEGY

In any study, choices must be made: choices regarding the presentation of the question, the objective, the actual organisation of the study and the instruments. These choices, however, are never obvious. In practice, the researcher always runs up against limitations. The degree of freedom available to him or her is not without restrictions. Time, money and access to the research subjects (in our case the companies) can form significant limitations. There is, however, too little recognition of this in the 'academic discourses' and in the instructions usually encountered in the respected manuals on research methods (Buchanan et al., 1988). Such manuals (cf. Hertog and Sluijs, 1995, p. 27) 'assume that research objectives will be clearly formulated and that organisations will be willing to co-operate. Moreover, they ignore the fact that research costs time and money'. The organisational researcher does not work in an ivory tower: researchers always function within the same social reality as their subjects. In practice, then, the researcher, and certainly the organisational researcher, will have to strike a balance between the desirable and the attainable (Box 3.1). The process is fraught with conflicts and dilemmas, and both fundamental and opportunistic choices must be made. Or, as Homans (1950) puts it: 'Methodology is a matter of strategy, not of morals. There are neither good nor bad methods but only particular circumstances in reaching objectives on the way to a distinct goal'. With these words (cited in Grunow, 1995, p. 95) Homans had made it clear as early as 1950 that the most important criterion for judging methods in the social sciences was the objective aimed at in a given situation. Homan's position (Grunow, 1995, p. 93) lies between two extremes: total relativism ('anything goes') and belief in 'one true and acceptable method'. In Homan's view, methodology in the social sciences is not a catechism but an individual and collective learning process based on experiences with certain methodological choices in the practice of research

(Hertog and Sluijs, 1995). The strategy followed in this project clearly has the nature of such a compromise. In the sense that the development of knowledge in the course of the project is considered to be a significant step in a knowledge enterprise (ibid.).

BOX 3.1 COMPROMISES

'Fieldwork is permeated with the conflict between what is theoretically desirable on the one hand and what is practically possible on the other. It is desirable to ensure representativeness in the sample, uniformity of interview procedures, adequate data collection across the range of topics to be explored, and so on. But the members of the organisation block access to information, constrain the time allowed for interviews, go on holiday, and join other organisations in the middle of your unfinished study. In the conflict between the desirable and the possible, the possible always wins. So whatever carefully constructed views the researcher has of the nature of social science research, of the process of theory development, of data collection methods, or of the status of different types of data, those views are constantly compromised by the practical realities, opportunities and constraints presented by organisational research.' (Buchanan et al., 1988, pp. 53–54, cited in Hertog and Sluijs, 1995, p. 27)

To describe the field of tension between relevance and innovation on the one hand and scientific accountability and soundness in the organisational sciences on the other, Donald Schön (1984, quoted in Kaplan, 1986b) rather cynically compares the environment of the organisational researcher to a swamp (Box 3.2).

It is a fundamental choice: the security of the beaten path (the high ground in the metaphor) or the adventure of exploration (which means entering the swamp). This research project had to confront this dilemma as well. And there is a need for such adventures in organisational science. As Daft and Lewin (1993, p. i; cited in Hertog and Sluijs, 1995, p. 19) warn: 'We are concerned that organisational theory is in danger of becoming isolated and irrelevant to the leading emergence of new paradigms'. They perceive a huge gap between the stormy developments facing companies and other organisations and the lethargic, inert manner in which the American organisational research 'mainstream' responds to such developments. Meyer et al. (1993) as well as Hertog and Sluijs (1995, p. 19) subscribe to this criticism and argue in favour of a holistic approach in which phenomena will be studied in the contexts within which they are embedded. From this point of view, it is the social system as a whole which gives meaning to the various elements which form part of it, and these elements cannot be understood in isolation from that whole. We are attempting to follow a similar holistic approach in this book.

BOX 3.2 A METAPHOR: THE RESEARCH FIELD AS A SWAMP

'Rigorous professional practice is conceived of as essentially technical. Its rigor depends on the use of describable, testable, replicable techniques derived from scientific research, based on knowledge that is objective, consensual, cumulative, and convergent' (Schön, 1984).

Unfortunately, as Schön observes: 'In the varied typography of professional practice, there is a high, hard ground which overlooks a swamp. On the high ground, manageable problems lend themselves to solutions through the use of research-based theory and technique. In the swampy lowlands, problems are messy and confusing and incapable of technical solution. The irony of this situation is that the problems of the high ground tend to be relatively unimportant to individuals or to society at large, however great their technical interest may be, while in the swamp lie the problems of greatest human concern. The practitioner is confronted with a choice. Shall he remain on the high ground where he can solve relatively unimportant problems according to his standards of rigor, or shall he descend to the swamp of important problems and nonrigorous inquiry?

Some researchers have continued to develop formal models for use in problems of high complexity and uncertainty, quite undeterred by the troubles incurred whenever a serious attempt is made to put such models into practice. They pursue an agenda driven by evolving questions of modelling theory and techniques, increasingly divergent from the context of actual practice.

A professional who really tried to confine his practice to the rigorous applications of research-based techniques would find not

> only that he could not work on the most important problems but
> that he could not practice in the real world at all' (ibid.).

But of course the scientific criteria of controllability also apply to studies
which are methodologically somewhat further from the beaten path. It is
important for researchers to make all their choices explicit, especially when
they make themselves vulnerable by moving away from the beaten path. For
this reason, the research design is of great importance. By this we mean the
coherent whole of arguments and assumptions which forms the basis for the
research concept, data collection, data analysis and reporting the research
results (Hertog and Sluijs, 1995, p. 244). The design represents the strategy
and the actual course of events, as well as, primarily, the logic which links
questions and answers. It ensures 'closure', that is, that the project will have
a definite beginning and end. It must be possible to evaluate that complete
whole on its own terms. 'Complete' means that questions and objectives
have been clearly formulated, the approach is described and justified,
analysis is carried out and conclusions are drawn.

A study can also form part of a greater whole. This is certainly true of the
present study, which forms part of a formal research programme.
Programmed research makes it possible to achieve objectives beyond the
scope of individual projects (Hoesel, 1985). Researchers can connect up
directly with previous steps and provide ammunition for new steps. This
means that not all steps of the empirical cycle need be gone through in any
one project. One project may be primarily exploratory, while another is
integrative and yet another might be aimed at verification of knowledge.
Combinations are also possible: current insights can be tested while a new
theory is developed on the basis of empirical observation. This project has
been set up and implemented on the basis of this point of view. It builds on a
qualitative, exploratory study[1] (project 1 in Figure 3.1) and the current
literature. The first project included a literature study and a qualitative study
of a number of companies in the processing industry. On this basis (see
Chapter 4), a 'provisional' working model was made: not a real theory, but a
temporary construct. Moreover, this earlier study provided the basis for the
development of a wide-ranging instrument which is useful in two ways:

- to test current insights (or rather, 'conjectures') on the basis of
  hypotheses, particularly when critical success factors are concerned. This
  is also the policy objective of this study, namely to determine whether the
  concepts concerning critical success factors which originated from
  previous studies could be applied in a broad range of industries and

service sectors to understand and improve the innovative success of firms (see section 2.4);
- to develop a general (non-sector-specific) theoretical model in which critical success factors are incorporated in a model of management competencies (this is the central theoretical and scientific objective).

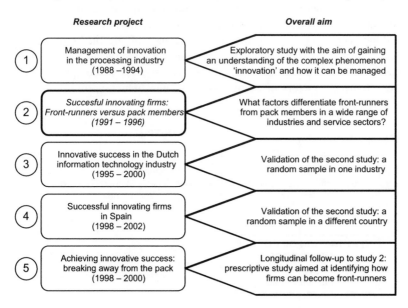

Figure 3.1 The incorporation of the project in a research programme

The second research project (on which this book reports) has thus been set up to serve both policy and scientific objectives. The scale of the project budget made it possible to depart from the beaten path and thus include not only verification of current insights but also the development of new ideas and instruments. In this sense, this project is intended to both verify and develop theory. The research programme does not stop there, however. In subsequent steps (projects 3 and 4), the emphasis will be placed on verification (and cross-validation). The objectives of these projects are thus to further refine the theoretical framework, to verify hypotheses which result from this study and to conduct cross-validations in different populations. In these more deductive follow-up studies, the validity of the bases for interpretation of random samples will be tested by means of an updated questionnaire. But new exploration (project 5) will also have a place, particularly when the time comes to delineate the dynamics of company development. It will then be a matter of acquiring a better grasp of the 'how' questions. How does a firm become a front-runner and maintain that position? Longitudinal research will thus be unavoidable here. Figures 3.1 and 3.2 show both the underlying logic and the actual relationships between projects in the programme.

The fact that this study is intended to generate theory does not imply that we started from scratch theoretically. When we set out to study the success of innovative firms, there was already a wide range of literature on the subject of innovation. However, as was argued in Chapter 2, most of the research has been done sector-wise, and while the major part of the literature covers problems related to the management of innovation projects, only a few studies have focused on achieving innovative success at company level. Studies of success factors of innovative companies which are not industry-specific are even more rare.

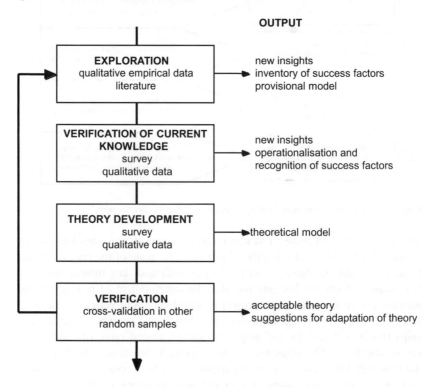

Figure 3.2 The logic underlying the relationships between the projects

Based on preliminary research and extensive study of the literature, a 'provisional' theoretical framework was thus developed on the basis of which a number of verifiable hypotheses could be formulated. These hypotheses, which will be presented in Chapter 4, can be regarded as the building blocks on which the theory is generated. Furthermore, in Chapters 5-8 a few detailed hypotheses will be formulated to test the findings of earlier research. Since the present study (project 2) is by nature an inductive, theory-generating study, the strategy has been consciously focused on first conducting a wide-ranging investigation (many subjects, many questions and a large number of sectors) into the reasons for innovative success which can be discerned in a large number of sectors.

3.2   RESEARCH DESIGN

3.2.1   Considerations and strategic choices

The study thus has a twofold research objective:
1. To identify (and/or verify) non-sector-specific factors which distinguish successfully innovating companies (front-runners) from companies which are only average for the sector in terms of innovative success (pack members). The emphasis here is on identifying factors which can be influenced by management;
2. To provide building blocks for a theoretical (competency) model incorporating these factors.

As noted earlier, a number of considerations were important for the formulation of the research structure:

- the present study is *at company level*, not project level;
- the study is *non-sector-specific*;
- as indicated in Chapter 2, we assume that differences in innovative success between companies can be explained by *differences in managerial competencies* (routines, skills and learning experiences). The question is how to identify such competencies in a wide range of sectors. The challenge is to single out success factors which are under the sphere of influence of the company: after all, we are not looking for success factors which result from environmental circumstances;
- the objective is to provide as *comprehensive* a picture as possible of the factors which explain innovative success. This implies that we shall be looking for connections between groups of variables in order to make it possible to integrate theoretical insights;
- a major part of the study is concerned with *theory development*. From the literature (for example Glaser and Strauss, 1967; Geertz, 1973), we learn that theory development requires a redundancy of data;

- to a limited extent, the study is *also intended to verify hypotheses*;
- the differences between front-runners and pack members will be measured among the better performing companies. After all, we are *not interested in the 'poorly' innovating companies*. The study is focused on providing companies with a stimulus to improvement;
- since this is a commissioned study, with a clear policy and company-relevant objective, the desires of the sponsor must be taken thoroughly into account. Among other things, this means that the *terms and concepts employed must also be recognisable and useful for managers*;
- 'Innovative success' must be made quantifiable in some manner. This means that *instrument development* must also be possible.

These considerations imply that a number of choices must be made. In the process, of course, account must always be taken of practical limitations such as the available time and research capacity, the research budget, the willingness of companies to participate, and the practical restrictions inherent in data collection (after all, not everything is measurable, often because the company does not have the data). Field research is always a compromise between what is desirable and what is possible. Consideration of the desirable and the possible has led to the following strategic choices.

Intensive data collection
Since there was no comprehensive list of innovation success factors at company level when the study began, we had to develop one ourselves. One of the objectives was thus to take the theory a step further on the basis of quantitative data. This, however, implies that a good deal of measurement must be carried out. For this reason, we decided on intensive data collection, also known in the literature of methodology as the 'trawler approach' (Daft and Lewin, 1993). We have thus attempted to build up a very comprehensive database which can be consulted during the process of formulating theory. This also makes it possible to look for alternative explanations for results. It demands a research set-up which enables us to carry out in-depth research at each company. Of course, the number of companies which can be investigated in this manner is limited.

Emphasis on qualitative as well as quantitative data
We do not believe that knowledge about factors underlying innovative companies can be developed by collecting or measuring 'hard figures' (such as R&D expenses, financial figures, throughput times, number of innovations) alone. The knowledge we are searching for is aimed rather at gaining insight into the meaning of these figures and the more fuzzy organisational processes and characteristics which allow a company to be

successful over an extended period of time. For that reason, a considerable amount of qualitative data must also be collected. Such qualitative data serve a number of objectives. First, they contribute to the wealth of material which can be investigated and which can be of help in explaining unexpected or contradicting findings. Second, they are also a means of triangulation[2], as the story behind the figures tells something of the validity. Third, qualitative data provide the building blocks and ideas to build new theories. And last, they provide a way of tracking down more dynamic elements. After all, the story behind a cross on a questionnaire has much more to say than the cross itself.

Matched pair comparison
When studying innovative success in a wide variety of sectors, basically three quantitative research designs are possible. In the first place, one can decide to carry out a large random sample of hundreds of companies. As the aim is to find non-sector-specific success factors, the influence of the sector has to be eliminated from or minimised in the analysis. This requires a large random sample. However, such a study would become very costly as the large number of factors which must be measured and the qualitative focus requires face to face interviews. Such large-scale data collection (and even redundancy) is necessary because we want to integrate current insights into critical success factors in a new theoretical model. So, with this methodology we would rapidly run up against budgetary and capacity restrictions.

An alternative is to work with random samples in a limited number of sectors. But if we really want to arrive at non-sector-specific results, this implies that some 30 randomly selected companies in various sectors must be involved. This will also lead to budgetary problems. In addition, based on political arguments; the sponsor (the Dutch Ministry of Economic Affairs) requested that at least 25 sectors be involved in the study. The political climate in the Netherlands at that time prevented sector-specific studies to a large extent. For the researchers, this limitation proved to be a challenge as it was not sure whether it could be done. Within the financial resources available to us, we thus had to choose between a wide-ranging study (many companies per sector, but little information per company) and an in-depth study (a few companies per sector, but those would be investigated in depth). We opted for a third alternative and selected a method which allowed us to investigate in depth.

This method is called the matched pairs design. In this design, sampling is performed on the basis of the dependent variable, innovation success, with one firm classified as front-runner and its twin as pack member. For each sector, a matched pair is selected (two companies which have a number of characteristics in common and which match with respect to a number of criteria but differ in terms of innovative success). By means of matching, environmental factors can be eliminated to a significant extent, since they are more or less the same for both companies, and the explanation for success can be sought in the differences in internal routines and competencies. The correspondences between the companies which form matched pairs will be sought in features which might be considered explanatory factors but whose influence will be expressly left out of the analysis (the so-called matching criteria):

- Competing for the same market;
- sales volume;
- ownership structure;
- solvency;
- and technological function.

The two companies in each pair will be matched in terms of these independent variables. We shall go into this more deeply in section 3.4. This method is a practical way of identifying general critical success factors in a limited number of companies by means of research and has been much employed in the literature; several studies in strategic research have proven the value of this research design for management success factors (Rothwell, 1972, 1974; Rothwell et al., 1974; Cooper, 1986; Schreuder et al., 1991). Moreover, the sponsor has had good experiences with this method in two other studies (Zwan et al., 1987; Schreuder et al., 1991). Besides its charms, however, this research design also has some pitfalls of which we are definitely aware and which must be taken thoroughly into account in the further development of the research design. I shall return to this point later in this chapter. As the matched pair design is the dominant characteristic of the research design, more attention will be paid to it in the following sections.

Theory-generating research
A selective sample is inherent in a matched pair comparison. Research subjects are not picked at random but selected consciously in advance. This need not pose any problems, since we are concerned here with theory-generating research. Theory-generating research endeavours to establish a theoretical generalisation, not a statistical one. So cases are chosen for theoretical, not statistical reasons (Glaser and Strauss, 1967) and therefore do not need to be sampled randomly (Eisenhardt, 1989, p. 537; Yin, 1984). In theory-generating research, findings are generalised to theory, analogous to the way in which scientists generalise from experimental research to theory[3] (Yin, 1984, p. 39). In this type of research, more importance is attached to the diversity of phenomena rather than the frequency with which a given

phenomenon occurs (Hertog and Sluijs, 1995), and a selective sample is often more suitable than a random sample.

However, the researcher does have to be aware of the possibility that the selection may include a bias. Subsequent statistical control and cross-validation with another sample is therefore essential. In this manner, the extent to which a bias may be present can be examined. The risk of bias is only great when a *systematic* bias is concerned. This is unlikely in our study: first, a different assessor was used for each sector; and second, we were able to check the process in retrospect. So, there is no reason to assume that a systematic bias has occurred in the selection of companies. It is not the case that we are now going to construct an entirely new theory or start from scratch. As described in Chapters 2 and 3, we began by provisionally basing ourselves on a simple theory, a temporary construct comparable to scaffolding on a construction site. The original research questions serve as a directive to lead us in this study. The research question (Strauss and Corbin, 1990, p. 39):

> gets the researcher started and helps him or her to stay focused throughout the research project. ... Then, through analysis of the data, which begins with the first collection of data (the first interview or observation), the process of refining and specifying the question will begin.

The final design is thus a hybrid form embodying both inductive and deductive elements. In addition to the development of a theory, after all, a number of hypotheses put forward in the current literature are to be tested.

No poorly performing companies (laggards) in the sample

As has been said, we decided to conduct a comparison between front-runners and *pack members*. We did not make things easy for ourselves by doing so. After all, much greater differences would come to light in a comparison between front-runners and laggards. Yet one may question the practical utility of a comparison with manifestly inferior companies which can't get anything right. Comparing the front-runners to the pack members, however, will yield results which can be useful to a wide range of companies. Pack members, after all, often have what it takes to develop into front-runners. This is much less true of laggards. Moreover, a comparison between front-runners and laggards would lead to recommendations for good management in general rather than good innovation management. The major drawback to a front-runner/pack comparison is that the differences will be smaller, and this is something we must take thoroughly into account.

Sample selection by outside informants

The success of the selected design depends on the criterion used to distribute the companies into front-runners and pack members respectively. It is of great importance to utilise an external criterion. The researchers themselves should not distribute the companies into front-runners and pack members in advance, otherwise they will run the risk of biased results. In addition, an external criterion makes it possible to test one's basic principles against field data. One then makes use of a different yardstick for the criterion which, ideally, will result in the same classification.

A basic problem in management research in general and innovation management in particular is that many aspects cannot be measured with quantifiable or so-called hard objective measures because of data unavailability or the lack of clearly defined, generally usable measures. In such circumstances, researchers usually resort to using informed practitioners and observers (Snow and Hambrick, 1980), who may come either from within the company being studied (inside informants) or from its environment (outside informants). While the former source of information is used extensively in academic literature, the latter has been relatively neglected. Increasingly, however, researchers are advocating the use of outside informants. Harrigan (1983), for instance, encourages researchers to conduct field interviews with outside informants to enhance understanding of firms and their competitors' strategies. Chen et al. (1993) reviewed the use of outside informants in strategy research reported in five major journals between 1986 and 1991. Of the 141 studies reviewed, 47 used academics, 11 used security analysts, 10 used consultants, 3 used suppliers, 11 used buyers, 37 used executives from outside the local firms, and 22 used other industry stakeholders. We also resorted to external informants (industry experts) in selecting front-runners and pack members. The study by Chen et al. (1993) further demonstrated that external industry experts can be a reliable and accurate source of strategic information. This study, executed among 100 industry participants as outside informants on the airline industry and 44 senior airline executives (as inside informants for the purpose of verifying validity and reliability), was aimed at measuring various strategic attributes of certain competitive moves in the airline industry. The study showed that informants in each of the four groups of outside informants studied (security analysts, consultants, stakeholders and academics) manifested high interrater reliability. Of the outsiders, analysts were the most accurate and were highly reliable, while academics were highly reliable and as accurate as consultants and stakeholders.

Intersubjectivity

Much organisational research is based purely on managers' judgements about themselves, on a subjective picture of their own situation. We are faced with this problem as well. Such intersubjectivity cannot be expelled from this type of research, but it can be reined in. To keep this intersubjectivity ('bias') under control, the following choices have been made:

- the questions in the precoded list are formulated as much as possible in concrete, descriptive (not evaluative) terms which are in keeping with the dominant 'overt' behaviour (routines, things as they occur);
- exact measurements are used where possible;
- diverse sources are utilised (four managers and external informants);
- diverse methods are employed (questionnaires, open interviews, documents).

3.2.2 Cross-case survey

This study aims at developing theory and testing state-of-the art knowledge about critical success factors. For obtaining answers to the basic research question ('What distinguishes front-runners from pack members?'), the survey appears to be a suitable method (Hertog and Sluijs, 1995). As a research design, the survey is appropriate to give answers to 'what' questions (Yin, 1990) which have external validity. The word 'design' is important in this respect. The survey is not regarded in this study as a method of data collection, but rather as a 'research plan'. Nachmias and Nachmias (1976; cited in Yin, 1990, pp. 28–29) define the design as 'a plan that guides the investigator in the process of collecting, analysing, and interpreting observations'. This definition implies that the researcher is not restricted in his or her study to the use of a questionnaire (as in the narrow definition of the survey), but can employ a wide range of instruments. Apart from the questionnaire, data can also be derived from archival records, observations, content analysis, and so on. Furthermore, the unit of analysis becomes a more appropriate identification of the object being studied than the respondent (Hertog and Sluijs, 1995), since one case may be based on data from multiple respondents and sources. In this we follow the definition of the survey as research design given by Marsh (1982, p. 6):

> An investigation where:
> - systematic measurements are made over a series yielding a rectangular of data;
> - the variables in the matrix are analysed to see if they show any patterns;
> - the subject matter is social.

But the survey design also has its limitations. Two of these are relevant within the framework of this study. The survey design allows little attention to be paid to the context within which a phenomenon takes place. Moreover, it provides only a glimpse of a certain moment in time. It is static, making it difficult to detect causal relationships and explain the observed phenomena in terms of processes or chains of events (Hertog and Sluijs, 1995). The result is that it is practically impossible to answer the 'how' questions (Yin, 1984), since this demands in-depth insight into the context in which the firms operate and the processes with which they are involved. Detecting causal relationships requires insight into path dependencies and possible alternative explanations. In order to get a better grip on alternative explanations, redundancy of data and a combination of qualitative and quantitative insights are essential. But this approach calls for a compromise, since it limits the size and randomness of the sample.

The case study method is more appropriate for studying the elusive phenomena of innovative success (Yin, 1984; Hertog and Sluijs, 1995). As Yin (1984) states: 'In general, case studies are the preferred research design when 'how' or 'why' questions are being posed, when the investigator has little control over events, and when the focus is on a contemporary phenomenon within some real-life context.' The researcher can then devote attention to the context within which the process of change takes place. Case studies tell a story and place events in perspective along a time path. Therefore, they are very useful in answering 'how' and 'why' questions. Although case studies are often a rich source of ideas (hypotheses) about various factors involved, they can be problematic when used to test these hypotheses. Yin (1990) argues that case studies can be used to test theory and generalise from data when a theoretical rather than a statistical perspective is followed. The case is then regarded as a 'natural experiment' and it is tested whether the theoretical expectations hold true for the case history. The risk remains, however, that the selection of the case(s) studied will be based on a strong bias. Cases are never selected in the dark. We almost always know the 'clue' in the story beforehand. This argument has even more weight when the objective (as in our study) is to identify success factors which are valid across a wide range of sectors.

Given the constraints in time and resources, the researcher's dilemma in this respect (Larsson, 1993) might be more pragmatic than paradigmatic: either sacrificing the richness of information to increase the number of cases in the sample or the other way around. This dilemma has inspired researchers to develop new designs which combine the strong points of both the case study and the survey. Different designations are used for these approaches: 'hybrid form' or 'case sampling' (Blanck and Turner, 1987), 'case meta-analysis' (Bullock and Tubbs, 1987) 'case survey' (Larsson, 1993), and 'dual methodologies' (Leonard-Barton, 1990). One of these alternatives, the 'case survey', is described in detail by Larsson (1993). Here the database is built on a collection of case descriptions concerning a specific issue which are available in the formal or 'grey' literature. These case descriptions form the basis of a meta-analysis in which a coding scheme is used for systematic conversion of the qualitative case descriptions into quantified variables. The coding is done by multiple raters (which makes it possible to measure interrater reliability). Finally; these coded data are subjected to statistical analysis. Larsson (1993, p. 1516) argues that such a design is especially suitable when: (a) case studies dominate an area of research; (b) the unit of analysis is the organisation; (c) a broad range of conditions is involved; and (d) an experimental design is impossible. However, this design has also a number of serious limitations. Most importantly, the researcher is always dependent on the availability of case descriptions in literature. This has four main consequences:

1. There might be a bias in the sampling of available cases, for example because success stories are far more easy to publish than failures.
2. In most studies there are considerable gaps between the times at which the described phenomena, the retrospective analysis and the final meta-analysis take place.
3. The researcher has no influence over the content and format of the data input. This limitation is threefold. In the first place, the coding implies a strong reduction of the variety of meaningful case data. In the second place, the researcher is very restricted in determining the variables he or she regards as relevant or 'interesting'. And last, there is no possibility of validating the reliability of the case data themselves.
4. This form of case survey can be used for theory testing, but due to its reductional slant it is not well-suited for theory development.

Glick et al. (1990) tried to remedy these limitations to some extent in their longitudinal studies of 'Changes in Organisation Design and Effectiveness' ('CODE'). In this study, the researchers built a data set of more than 120 organisations selected on the basis of their own *a priori* criteria. Data gathering took place by means of repeated semi-structured interviews and questionnaires. Qualitative data (the retrospective event histories) were coded by means of an *a priori* coding instrument. This approach implies a considerable investment in data gathering. The advantage is that the researchers are 'in control' of their own study, rather than being dependent on the field work and case descriptions of others.

Given the dual objective of this study (theory development and theory testing) and our interest in both 'factors' ('what' questions) and processes ('why' and 'how' questions), the choice was made of a similar approach, the *cross-case survey*. The cross-case survey in this study is based on the following elements:

- the organisation is the unit of analysis;
- the sample is constructed on the basis of *a priori* criteria and taken from a population of organisations defined in an *a priori* manner;
- data for each organisation are collected from a variety of sources: managers (multiple respondents per firm), industry experts, and documents (annual reports, newspaper articles, and so on);
- data are gathered via multiple instruments: semi-structured interviews, precoded questionnaires and content analysis;
- the final database has two parts: a data matrix composed of the quantitative data and a collection of company-based case descriptions;
- the data analysis follows an iterative path between the quantitative and the qualitative data.

This combination of quantitative and qualitative data can be highly synergetic. The redundancy in the large combined database creates the right conditions for grounded theory. New ideas can be almost continuously developed and tested. Quantitative evidence can indicate relationships which may not be salient to the researcher (Eisenhardt, 1989, p. 538). Furthermore, the qualitative data are useful for understanding the rationale underlying relationships which emerge from the quantitative data. In fact, it can be questioned whether a survey alone is adequate to cover the significant elements of social behaviour (Vaus, 1991, p. 331). After all, a correlation tells us nothing about the direction or cause of the relationship. The qualitative data can also serve as a way of enhancing the translatability (Dunn and Swierczek, 1977) of the study results for practitioners. Also, the qualitative data are especially useful in finding a posteriori explanations ('why' questions) for unexpected results of the quantitative analysis. Finally, the alternation between qualitative and quantitative data facilitates triangulation and increases the validity of the findings (Eisenhardt, 1989, p. 538).

Of course, the cross-case survey design also has its risks. Like all research designs in the social sciences, it must be regarded as a compromise, in this case a compromise (McGrath et al., 1982) between general applicability, richness of information and practical relevance. McGrath et al. argue that in any design the researcher can do no more than satisfy two of these three conditions at the expense of the third. A cynical colleague might object that this study has made sacrifices with regard to all three conditions. From a quantitative perspective one might remark that our sponsor's limited resources and objectives have certainly not resulted in an ideal sample. In contrast, the qualitative researcher might determine that this large sample of cases only offers the possibility of 'thin' rather than 'thick' description. Finally, the practitioner might object to the academic form of this book. We have become increasingly aware of these risks in the course of the study. It is very difficult indeed to satisfy the three conditions mentioned above within the scope of one study or one book. Some of the risks can be dealt with outside the framework of this publication. We refer in this respect to the cross-validation studies which are underway and to the practice-orientated publications we wrote parallel to this thesis. At last a small number of cases have been subject to secondary data gathering and analysis (see, for example, Hertog and Huizenga, 1997). However, the possibility to alternate between quantitative and qualitative case data remains crucial in the present design. The values in our data matrix were not simply numbers. At all stages we knew 'the story behind the numbers'. To date, we have no clear-cut methods available for this iterative process. To a large extent, we have had to rely on our intuition and heuristics. In developing such methods, we are following the advice of Covalski and Dirsmith (1990, p. 566) and *doing* (qualitative) field research rather than merely *talking* about it.

3.3   THE CRITERION VARIABLE: INNOVATION SUCCESS

In order to make a distinction between successful innovators (front-runners) and less successful innovators (pack members), an unambiguous criterion is necessary. This requires a measurement of input/output effectiveness which is *as unambiguous as possible* and *not purely dependent on the subjective judgements of the managers themselves*. Finding such a yardstick is not so easy. This is a problem of innovation research in general which every researcher soon encounters. An extra difficulty in the present study is that we are looking for an external criterion for measuring effectiveness that is also *generally applicable*. After all, we want to compare front-runners and pack members from a broad range of sectors. But input, output and innovation are very difficult to measure when the study covers different sectors. Since organisations play 'different ball games', the importance of many criteria is industry-specific. Yet one must be able to compare innovative effectiveness in a science-driven sector such as the pharmaceutical industry with innovative effectiveness in a bank. In fact, the criterion for innovative success encompasses various dimensions and varies according to company and to phase in the company's life-cycle. This ambiguity of definition makes it very difficult to conduct a comparison between companies and sectors. An unambiguous, non-sector-specific instrument for measuring innovative success is thus difficult to achieve and, in fact, would require a study of its own.

Our design requires a criterion for success which allows us to select the front-runners and pack members to be studied. However, there was no database available for this purpose, nor did we want the company managers to be the sole judges of their firms' innovative success. The process of finding a good criterion is also complicated by the fact that we are looking for long-term yardsticks which can *measure innovative effectiveness and innovative success at organisational level*. The literature does offer a great many yardsticks for measuring innovative effectiveness at project level, but the object of this study is not the individual innovation project but the innovative organisation. This calls for a criterion which can enable us to distinguish successfully innovating companies from less successfully innovating companies. Innovative success is a multiple criterion which, moreover, is embedded in a great many other factors. Due to the interweaving of the functions related to the phenomenon of innovation, finding such a criterion is considerably more difficult than finding yardsticks to measure the effectiveness of the whole organisation (Zwan et al., 1987; Schreuder et al., 1991). Under these circumstances, it is an accepted option to use external experts as judges.

There are thus still no unambiguous and directly usable criteria in the literature which measure the innovative success of companies for a broad range of manufacturing and service sectors. The available criteria are too general, too detailed or sector-specific. While the resulting problems are definitely solvable, they demand an intensive approach.

Considering the complexity of the material and the inadequate, limited public information available concerning the innovative success of individual companies, the logical option is to employ a broad definition of innovative success and engage external judges to make a selection on the basis of their own, heuristic criteria[4] (see Tversky and Kahneman, 1974; Hershey and Schoemaker, 1989).[5] This is the alternative we have chosen in this study. For each sector, the front-runner and the pack member have been selected by an external industry expert on the basis of his or her own heuristic criteria. In addition, this procedure allows us to perform a statistical check on the basis of the data we shall gather within the companies. In order to make the heuristic process more explicit, the experts were presented with a vignette indicating what the researchers wanted to measure in the companies (Box 3.3).

We assume that firms innovate with the intent of bringing to the market new products which satisfy some customer need. True, many innovations have been launched that failed to enjoy market acceptance, but such innovations would not classify the firm as a *successfully* innovating firm. We stress the fact that our focus is on innovativeness which yields positive returns or the accumulation of permanent intangible assets that confer a competitive advantage. We are not simply interested in a firm's innovation scorecard; rather, we are attempting to identify the stream of successfully introduced innovations which make the firm excellent as a whole. The manner in which this selection was effected is described in subsection 3.4.2.

BOX 3.3 VIGNETTE SHOWN TO OUTSIDE INFORMANTS WHO PERFORMED THE SELECTION

Innovation: Renewal of services, products, processes and organisation. New means 'new to the company'.

Successful innovation: The economic exploitation of innovation. This can be achieved in different ways, such as increasing turnover by introducing new products, increased productivity through the introduc-tion of new processes, enhancement of services, and so on.

3.4   SAMPLING

It is not only high-tech firms which add new products or services or improve their production processes. This happens in most firms and is equally applicable to 'mature' businesses such as automobiles, banking, auditing and cattle feed. Therefore, it was our intention that the study should not be limited to so-called high-tech organisations but should include low-tech and medium-tech firms as well. By including a wide range of industries, the present study can inform us about general success factors that are neither idiosyncratic to certain lines of technology nor to specific industries and markets.

The sampling was based on a nested design. First, a sample of 45 sectors (manufacturing and services) was taken. Second, a matched pair of firms was selected by industry experts within each sector. This resulted in a cross-section of the medium-sized and medium-tech Dutch trade and business community: manufacturing and service sector firms, ranging from insurance companies, banks and consultancies to chemical, electronic and metalworking companies. This sampling procedure will be discussed in more detail in the following two subsections.

3.4.1 Sampling of the sectors

The selection of sectors was based on data provided by the Dutch Central Bureau of Statistics (CBS). The CBS distinguishes nine industries, namely: agriculture and fisheries (SIC code 0); mining (1); industry (2/3); public utility companies (4); construction and contracting firms (5); commercial, hotel and restaurant sector, consumer article repair firms (6); transport, storage and communication companies (7); bank and insurance sector, business services (8) and other services (9). In consultation with the sponsor, it was decided to limit the study to the industries with SIC codes 0, 2, 3, 5, 6, 7 and 8. CBS has divided the industries into subgroups, which are subdivided in turn. The deeper one penetrates into the hierarchy, the better the companies meet the 'competing for the same market' criterion. On the other hand, however, it becomes increasingly difficult to find matched pairs, since the subpopulations become smaller and smaller. Previous matched pair research in the Netherlands (Schreuder et al., 1991) showed that the three- and four-digit levels are the most suitable for the Dutch situation. We thus decided to use the 'industrial group'-level (three-digit level) as the basis for selection. Consequently, a 'sector' is defined at the three-digit level. The selection of the sectors (at three-digit level) was done in two steps.

First, the two-digit industry classes (from which the sectors would be drawn randomly) were selected by means of a stratified random sample. The reason for the stratification was to ensure as heterogeneous a sample as possible. The stratification was done on the basis of the total turnover generated by the sector. We decided on turnover as the stratification criterion because this placed the emphasis more in the direction of companies with strong technological bases. If the 'importance' criterion had been the number of employees, emphasis would have been placed on more labour-intensive companies.

Second, each time an industrial class (two-digit level) was selected, a sector (three-digit level) within that class was randomly drawn (with the help of a random number generator).[6] To enable good pair matching, there was one constraint to the final selection of the sector, which was that it had to encompass at least ten companies (each with at least 50 employees). The reason for this constraint is that a sector has to include sufficient companies to provide a reasonable chance of forming a matched pair. To avoid making the sample too heterogeneous, we decided to reduce the size dependency to some extent by including only companies with more than 50 employees.[7] Furthermore, we decided to focus on well-established companies only, since other contingencies apply to starters or recently started companies. Concretely, this meant that only sectors with more than ten companies (with at least 50 employees) were included in the sample population.

The sectors in the final sector sample differed with respect to the relative emphasis placed on the process, the product or the type of innovations; and included both industrial and service sectors (Appendix 1 provides the final list of sectors studied).

3.4.2 Matching the pairs

In each of the selected sectors, a pair (consisting of a front-runner and a pack member) was formed. As was indicated in the previous subsection, we decided to rely on sector experts to perform the matching. Sector experts were invited on the basis of their thorough knowledge of the sector (at least five years of experience) and of the firms in that sector. For reasons of objectivity, we did not use sector experts who were employed by a firm in the sector. These people could generally be found in trade or industry associations, banks (security analysts, among others), sector-wise organised consultancy companies and, in rare cases, among sector suppliers or customers. To enable the sector experts to speak frankly about the firms, they were promised anonymity.

The matching was done by them (each for his or her own sector of expertise) in the course of a structured interview. The experts were first asked to describe how they would like to measure innovative success for their sector and which (heuristic) criteria they would apply. They were then asked to name one company which, according to their own criteria of innovative success, could be regarded as being very successful, the front-runner. They were then asked to name another company active in their sector which resembled the front-runner with respect to the matching criteria but which could be regarded as a good representative of the average company in that sector in terms of innovative success (the pack member). We specifically asked them to select an average company rather than a poorly performing company. The matching criteria we used were 'competing for the same market' and 'ownership structure' along with at least one of the following three: 'technological function', 'size (annual turnover)' and 'solvency', since these are believed to be variables which might explain differences in innovative success but which will not be studied as such in the current study. The experts were then asked to rank each company in terms of eleven predefined success criteria. In addition, they were also asked how much significance they attached to each of these predefined criteria as good criteria for innovative success for their sector.

As Table 3.1 shows, the matching of the firms, as indicated by the expert, was done reasonably well, so it is fairly safe to assume that the current data set is a matched pairs sample. The percentage of firms which did not match on one of the criteria ranged from 0 per cent on the 'competing for the same market' and 'ownership structure' criteria to 39 per cent on 'size'. At least this was the impression of the expert. However, the data did not show such great differences. We received data on turnover from both companies regarding only 12 complete pairs. These data showed that the average pack member had an annual turnover of 71 million Dutch guilders and the average front-runner had a turnover of 74 million Dutch guilders (see also Figure 3.3).

Table 3.1 Precision of match

Criterion	Match	No match	Unknown by expert
Competing for the same market	100	0	0
Ownership structure	84	0	16
Technological function	77	7	16
Size (annual turnover)	42	39	19
Solvency	23	10	68

Thus, in retrospect, we have indications that 'size' was also a good matching criteria. 'Solvency' did not appear to be a safe matching criterion since 68 per cent of the experts did not know whether the firms matched on this criterion or not. It is not a perfect match, but that is practically impossible. It is impossible to strive for a 'perfect' match, but we have tried to approach a perfect match as nearly as possible, for example, by asking the expert for a second or sometimes even a third pair.

It was then examined whether these firms had somewhat equal starting points one to ten years ago. The experts were asked to indicate, on the basis of their own success criteria or heuristics, whether they thought that the two firms could be said to have been equally successful in the past. According to the experts, in only 10 per cent of the cases were the two firms forming a matched pair not equally successful at some point in time between five to ten years ago. Firms forming the pairs did have such an equal starting point in 58 per cent of the sectors. In almost one-third of the cases, either the experts did not know whether these firms had an equal starting point or one of the firms did not exist in its current structure three years ago.

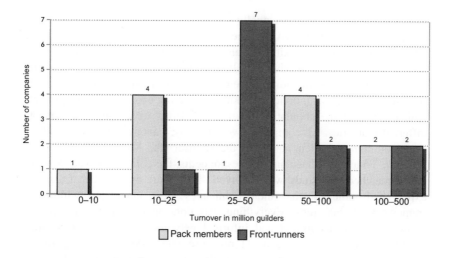

Figure 3.3 Turnover of 12 complete pairs (24 companies)

3.4.3 Unit of analysis

The unit of analysis of this study is the firm or, more specifically, the organisational unit, which at the minimum includes a full line of business supported by marketing, R&D (or product development) and manufacturing or service delivery. For most of the smaller companies, this meant that the unit of analysis was the company as a whole. In some of the larger (often multidivisional) companies, however, the unit of analysis was a business unit or other independent unit with its own marketing, R&D and operations. The variables are considered to be features of the company as a whole: which also implies that they can be related to structures and processes that can be recognised at lower levels (department or team level, for example). In this we follow most of the literature on critical success factors and organisational differences in innovation effectiveness (cf. Pennings et al., 1991).

3.5 PROCEDURES AND INSTRUMENTS

3.5.1 Access to companies

In principle, a matched pairs design is vulnerable in terms of response, since it requires both firms to participate. For this reason, and also to minimise the non-response bias, a lot of effort was put into encouraging participation, beginning with the invitation to participate, which was accompanied by a letter of recommendation personally signed by the Chairman of the umbrella employers' association (Algemene Werkgeversvereniging, AWV), the Chairman of the metalworking and electronics employers' association (FME) and the Director of the Technology Policy Directorate of the Dutch Ministry of Economic Affairs. In addition, the companies were promised copies of the final report as well as individual reports comparing each individual company with the rest of the population. In total, 90 firms were contacted from 45 different industries. All companies were promised anonymity. We would not even communicate their names to the sponsor (the Ministry of Economic Affairs). To avoid influencing the respondents, the companies were not told that a matched pair study was involved. They were only told that a sector expert had indicated that their companies were interesting within the framework of this study. Everything was done to persuade companies to participate. Various formal and informal networks were used. Telephone follow-ups were utilised to boost response rates. On average, up to ten attempts were made to reach firms which had not responded to our letter. In some cases, as many as 25 telephone calls were made to persuade a firm to participate.

About half of the firms which initially declined eventually decided to participate after several phone calls or presentations to their company. In all, 63 firms from a total of 35 industries responded positively (response rate of 70 per cent for the firms and 78 per cent for the sectors). A list of sectors studied is provided in Appendix 1. Due to non-response of one of the pair members in some sectors, it was not possible to form pairs in all sectors. Ultimately, the number of complete pairs was fewer than we would have liked. Subsequently, therefore, we dropped the matched pair design and decided to change to a less strict form, the matched sample design (Yin, 1984). In a matched sample design, the two groups are more or less comparable, not all pairs. In retrospect, this decision appears to have had no real influence on the contents of the conclusions. When only the complete pairs were included in the analysis the same results were obtained as when all companies were included in the analysis. Including more companies in the analysis, however, increased the significance and robustness of the findings.

3.5.2 Data collection

Data were collected from both primary and secondary sources. The primary sources were extensive interviews with four senior managers in each firm: the R&D manager; the marketing manager, the general manager or, if applicable, the CEO, and a 'liaison person'. The latter provided general information about the firm, its history and organisation. Each interview took about 1.5 hours on average. The interviews were conducted on the basis of a structured questionnaire developed by the research team. Discussion took place on the basis of open-ended questions concerning the firm's innovation strategy and the way in which the firm turned this strategy into action. Additionally, a list of precoded questions was presented. If any data were missing or unclear after the first visit, the company was requested by telephone to provide the appropriate data. A vast body of information concerning the firms' technological, organisational and marketing competencies was thus collected.

The interviews were conducted from August 1992 to May 1994. Firms were visited by teams of two researchers, generally composed of a senior researcher and a junior researcher. The senior would lead the interview (ask the questions, engage in discussion if necessary and point to inconsistencies in the answers), while the junior would record the answers. The interview teams were composed alternately from a multidisciplinary team of three senior researchers and five junior researchers. The results were recorded in a predesigned format which ensured that findings from different researchers would be easily comparable.

In addition to the primary data, secondary data were gathered from archives, annual reports, newspapers, business journals and the transcripts of the expert interviews. Furthermore, the general findings were discussed regularly with a steering committee comprised of representatives from the sponsor, the Dutch Ministry of Economic Affairs, senior representatives of organisations which facilitate, fund or advise on innovations and representatives from the academic community.

3.5.3 The questionnaire

Since there was no comprehensive list of innovation success factors at company level in existence at the time this study began, we had to develop one ourselves. After all, one of the objectives of the study was to take the theory a step further on the basis of quantitative data. As mentioned earlier, a redundancy of data is needed to develop a theory. This implies, however, that a considerable amount of measurement must be done. For this reason, an extensive questionnaire was created. To make the connection with the broad field of theories and hypotheses in the area of innovation management, it was also necessary to collect a great amount of qualitative data on each company. Among other things, this enabled us to track down the more dynamic aspects.

The questionnaire each firm was asked to answer was therefore quite extensive (65 open-ended questions and 486 precoded items per company). It was the intention of the researchers to collect as much relevant information from the firms as possible. The questionnaire originated from a systematic search in the literature, followed by a meta-planning session among the three senior researchers involved. The aim of this session was to generate as many potential explanations as possible for innovative success. These possible explanations were then grouped into 'phenomena', which in turn were translated into questions that might be asked to get a grip on these phenomena. In this respect, uncopyrighted questionnaires from earlier research were skimmed and new questions were formulated when none was available.

The draft questionnaire was then discussed in a broader group of researchers and the questions were considered in terms of appropriateness and usefulness. The steering committee provided some additional comments. After that, a final list of questions was composed which was then subdivided according to the four types of respondents to be interviewed. For each type of respondent (R&D managers, marketing managers, CEOs and liaison persons, see Appendix 2) a different questionnaire was used, although some questions appeared on more than one questionnaire or even on all four questionnaires. A pilot test of this questionnaire was done at two firms (one from industry and one from a service sector), after which the final questionnaire[8] was constructed. In order to achieve a triangulation of the data, some questions were first probed with an open question (unaided recall) and became more specific (aided recall) later in the interviews.

Questionnaire I: the liaison person
The liaison person provided the researchers with general information about the firm and the organisation of innovation processes in particular. The questionnaire consists of five major parts. The first part deals with the company profile and involves, among other things, questions related to the sector, the environment, the organisational structure and the characteristics of the products or services. The second part is the most extensive and deals with the organisation of the innovation processes. It contains questions related to the phases of the innovation processes; the tasks, responsibilities and authorities; and human resources management. The third part is devoted to internal and external co-operation and communication. The fourth and fifth parts of this questionnaire are relatively short and cover the financing of innovation projects and the experiences of the firm with quality assurance. In the smaller firms in the sample, only three persons were interviewed: the R&D manager, the marketing manager and the general manager. The liaison person questionnaire was then answered by one or two of the other respondents.

Questionnaire II: the marketing manager
The first part of this questionnaire deals with questions related to the firm's markets, marketing strategies and marketing capabilities. The second part

covers the generation and detection of ideas for new product development. Customer and supplier relations regarding innovation are among the topics asked about. The third part goes into the organisation of the innovation process. It includes questions about planning, selection and evaluation. The last part deals with internal and external co-operation and communication.

Questionnaire III: the R&D manager

The questionnaire for the R&D manager is the most extensive of all. Interviews with R&D managers therefore tended to be somewhat longer than with the other respondents. The first part of the questionnaire deals with the firm's R&D programme for product and process innovation. It includes questions related to the R&D programme goals and budget, influences of governmental policies, and career patterns within R&D. The second part covers the technological position of the firm and includes questions related to technology development, patents and sources for product innovations. The next section goes into the organisation of the innovation process and includes the same questions which are posed to the marketing manager on this subject as well as some additional questions. It is followed by a similar section on internal and external co-operation and communication. The final part of this questionnaire deals with process innovations. Insight is obtained into the sources for process innovations and the goals of process improvements as well as their results.

Questionnaire IV: the CEO

In the CEO interviews, a slightly different approach was adopted. The interview started with an open conversation about the firm's innovative efforts. Although the researchers had a list of open questions, these were rarely asked as explicitly as they were formulated on paper. Instead, clarifying or interview-steering questions were posed during the course of the interview. This procedure contributed significantly to an open atmosphere. At the end of the open interview, however, the researchers checked whether all 13 open questions had been answered. Unanswered questions were then posed explicitly. The open questions dealt with the meaning of innovation for the firm, its policies and strategies, the involvement of top management with innovation, the effectiveness of the innovative organisation, the financing philosophy, recent actions undertaken to improve innovative success and lessons learned from the past.

After the open interview, the CEO was asked to fill in a precoded questionnaire in the presence of the researchers. This questionnaire was subdivided into the following sections:

* priorities and company policy;

- company profile regarding innovation and innovative success;
- internal and external co-operation and communication.

In addition, the CEO was handed a short questionnaire related to financial data, R&D spending, annual turnover and profit, along with questions related to the firm's performance regarding innovation (speed of innovation, effectiveness of innovation, and so on). Since this questionnaire often required additional archival research, it was usually mailed back to the researchers within a month after the interview.

3.5.4 Variable types and statistical analyses

Most variables used in the study were measured on scales ranging from 1 to 5. Respondents were told that they had to regard the intervals between the consecutive values on the scale as equal. Most respondents did so, as can be deduced, for instance, from the fact that they sometimes gave halves instead of whole numbers. For ease of communication, however, explanations were given for the variable values 1, 2, 3, 4 and 5, such as 'not at all', 'hardly', and so on. In fact, although the scales were intended to be used as interval scales, there is a chance that some respondents may have used them as ordinal scales. This poses no problem for the analysis, however, since we have generally summed individual items into composite variables (for example, by summing various items or averaging the same items from various respondents in one company). In academic practice, it is widely accepted that total scores from individual ordinal items can be treated as interval scales (Nunnally and Bernstein, 1994), for instance, when the arithmetical mean score (which assumes equality of intervals) is computed. By summing items to obtain a total score, we are implicitly using a scaling model to convert data from a lower (ordinal) to a higher (interval) level of measurement. Nunnaly and Bernstein (1994, p. 16) argue that 'the results of summing (ordinal) item responses are usually indistinguishable from using more formal methods'.

More fundamentally, Stevens's (1946, 1951) representational theory, which argues that the sophistication of the quantitative procedures to be performed stands in relation to the type of measurement scale,[9] is increasingly being disputed. Michell (1986, p. 406) argues that 'the representational theory entails no prescriptions about the use of statistical procedures (as Stevens thought), nor does it imply anything about the meaningfulness of measurement statements (as current belief has it)'. Nunnaly and Bernstein (1994) argue that Stevens's position can easily become too narrow and counterproductive. In reference to Stevens (1946, 1951) and a range of textbook writers who base their ideas on Stevens's propositions, Gaito (1980, p. 364) argues, slightly irritated, that 'These writers do not read the statistical journal literature, inasmuch as a number of articles on this topic showed clearly that measurement scales are not related to statistical techniques'. He concludes that Stevens's conception is shown to be a fallacy. Michell (1986) describes two other traditions which he terms operational theory (Gaito, 1980; Bridgman, 1928) and classical theory (Rozeboom, 1966). Neither accepts Stevens's view that one must have achieved a particular level of measurement to perform a particular statistical operation. Operational theory implies that there is no relation between measurement scales and appropriate statistical procedures. Classical theory rejects Stevens's scale-type distinctions and does not prohibit the use of any statistical procedures with measurements (Michell, 1986, p. 406). Nunnaly and Bernstein (1994) argue that using presumably impermissible transformations usually makes little, if any, difference to the results of the most common analyses.

3.6   THE SAMPLE

3.6.1   Descriptive characteristics of the sample

The final sample of the main study included 24 matched pairs of front-runners and pack members and 15 incomplete pairs, of which ten were front-runners and five were pack members. We could say that, together, the participating firms constitute a cross-section of the Dutch manufacturing and trade sectors; with a relative overrepresentation of manufacturing sectors (as more than half of Dutch employment is in service sectors); 35 per cent of the sample operates in the services sector, 25 per cent supplies semi-finished products and 40 per cent produces end products (Figure 3.4).

Although the companies were generally very open in the interviews, they were less open in providing us with financial data. For instance, only 14 complete pairs of both front-runners and pack members provided us with data on their gross profit as a percentage of turnover. For these 14 complete pairs, front-runners appeared to have a significantly higher profit ratio than pack members: 7.7 per cent versus 4.3 per cent (Figure 3.5). Of this group, the average front-runner was about equal in size to the average pack member, with the average pack member having 357 people employed and the front-runner 333. In fact, for the whole population we can say that the firms studied are medium-sized. Data from the complete pairs who provided us with data on turnover showed that the average pack member had a turnover of 71 million Dutch guilders and the average front-runner just slightly more (74 million Dutch guilders).

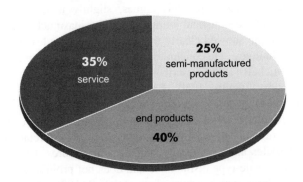

Figure 3.4 Composition of the random sample by product group (n = 63)

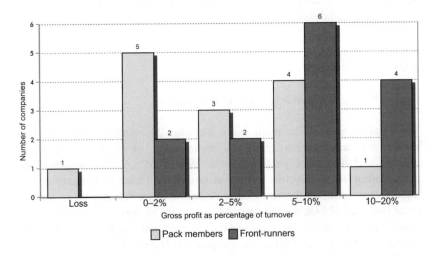

Figure 3.5   Gross profit as a percentage of turnover from 14 complete pairs (28 firms)

3.6.2   Treatment of non-response

We have tried to obtain the best possible insight into the reasons why some companies did not participate in the main study. These reasons for non-participation are often the same as those initially given by some companies from the response group, such as lack of time, an overabundance of requests to participate in scientific research, a recent reorganisation or company policy. Moreover, since about half of the companies which initially refused to participate were eventually persuaded, there is no reason to assume that the composition of the non-response group differs significantly from the remaining random sample. This assumption is reinforced by the fact that 11 companies which did participate in a follow-up study showed no significantly different results.[10] In any case, it must be noted that the front-runners were usually more willing to participate than the pack members. One might say that it seems as though the front-runners are more curious about the manner in which outsiders judge their companies than are the pack members.

3.6.3 The dependent variable

The dependent variable is the front-runner/pack member dichotomy, which is determined by the external expert. The variable only has two values: 0 if the firm is a pack member and 1 if it is a front-runner.

However, we did not want to rely only on this pre-set criterion, so have also constructed a second criterion by involving two other sources in addition to the expert judgement, namely the researchers and the company directors. The reasons for creating this extra variable were:

- to allow for some kind of triangulation, which increases the stability of the dependent variable if not all the pairs involved are complete;
- to allow for a greater distribution within the two groups of companies (front-runners and pack members);
- to create a (semi)-continuous scale, which permits certain statistical procedures not allowed by a dichotomous scale.

The semi-continuous variable is based on the following ratings:

1. *The expert rating* represents the score as furnished by the industry expert who selected the pairs of firms to be interviewed. This rating is 1 if the firm is a front-runner and 0 if it is a pack member.
2. *Researchers' judgement*: the research team collected both interview and archival information on each firm and generated its own rating (OURSCORE). Each firm was independently scored by the two researchers who visited the firm. A total of seven researchers visited firms in teams of various compositions. The research team rating represents overall innovative performance as rated by the two field researchers who visited a firm. On a five-point scale they might split between 1 (laggard), 2 (pack member), 3 (more than average pack member), 4 (catching up) and 5 (front-runner). The researchers were not put into fixed teams but visited the companies in changing compositions of one senior and one junior. It was therefore not possible to calculate

interrater reliability scores, but evidence of reliability could be found in the fact that the two interviewers showed 100 per cent agreement in indicating whether they regarded a firm as good to very well performing (scores 3, 4 and 5) or as mediocre to average performing (1, 2 or 3). The researchers, however, disagreed in 40 per cent of the cases with respect to the magnitude they attributed to the firm's innovative success (1, 2 or 3 resp. 3, 4 or 5). It is important to know that the researchers were aware of the expert judgements on the companies (front-runners or pack members) and that a great degree of correlation with the expert judgement can be expected. But since they (in contrast with the experts) were able to investigate different companies in different sectors, it was possible for the researchers to make further differentiations in the 0/1 dichotomy (pack member/front-runner). Front-runners in some sectors, for example, are head and shoulders above front-runners in other sectors when it comes to demonstrable innovative success. Naturally, the expert had no access to this information. The 'researcher's judgement' must thus be seen as a further differentiation of the expert judgement based on knowledge of a wide range of sectors more than an entirely independent assessment, which, however, is true in the following case.

3. *CEO / general manager judgement*: The general manager (or CEO) was asked to indicate how well the firm had been performing as compared with its competitors on a list of items indicating successful innovativeness. The possible answers were 'Front-runner' (5); 'Leading the pack' (4); 'Member of the pack' (3); 'Trailing the pack' (2); and 'Laggard' (1). The 12 items measured the success of the firm with respect to product innovations, process innovations, creativity, and so on. Since Cronbach's alpha of these 12 items was high (0.82), we decided to construct a scale ranging from 1 to 5 for these twelve items to indicate the average CEO's judgement on them (MEANCEO, being the average of the scores on the 12 items). Furthermore, the CEO was asked to give a grade from 1 to 5 indicating the overall success of the firm's innovative efforts. This grade was then averaged with the MEANCEO scale, thereby creating a variable (CEOSCORE) with scores ranging from 1 to 5 by means of which the CEO can give his firm an overall score for product and process innovative success. It is important to note that the CEO was unaware of the fact that this was a 'matched pair' study or that his company had been selected by an outside expert as being a front-runner or pack member.

The final computation of the dependent variable was done as follows. The judgements of the CEO (CEOSCORE) and the research team (OURSCORE) were recalculated so that they too ranged from 0 to 1 as did the expert judgements (EXPSCORE). All three judgements were then averaged into the final dependent variable: SUCCESS. In short, the SUCCESS variable was calculated as follows: SUCCESS = [EXPSCORE + ((CEOSCORE −1)/4) + ((OURSCORE −1)/4)]/3. It appeared after calculation that not one front-runner had a lower SUCCESS score than the best pack member. The SUCCESS variable can thus also be seen as a refinement of the front-runner/pack member dichotomy.

The resulting indicator, labelled 'Success' would then become a second criterion variable. As this new variable can be regarded as a scale derived from the three judgements, Cronbach's alpha was calculated, which proved to be sufficiently high (0.67) to validate the SUCCESS score. Since the intention of this success score was to encapsulate viewpoints from three different judges, it was expected that the alpha would not be too high. The CEO-derived scores, in particular, proved to be mavericks since the alpha rose to 0.72 if this variable were dropped. This small increase in alpha, however, did not justify losing the important and unique information in the CEO's judgement. Therefore, the variable was not dropped. The weak correlation of the score by the CEO with those of the experts and research team can to a great extent be attributed to the fact that many pack members regarded themselves as more successful than then they were regarded by the researchers, external experts and front-runner CEOs. After the research project had been conducted, this tendency to overestimate themselves proved to be typical of pack members and also an aspect which prevented some of them from changing. On the other hand, several front-runners underestimated their position as successfully innovating companies or their innovativeness.

To assess the convergence, possible criterion variables were correlated with each other. Table 3.2 provides the results of this triangulation. From this table we learn that SUCCESS is highly and significantly correlated with the scores generated by the research team and the expert, and moderately (but highly significantly) correlated with the more biased CEO-derived score. Based on this information, we might conclude that SUCCESS is a safe proxy for the dependent variable 'innovative success'.

Table 3.2 Correlation coefficients between the four measures of innovative success

	EXPSCORE	CEOSCORE	OURSCORE
SUCCESS	0.94**	0.45**	0.88**
EXPSCORE		0.19	0.73**
CEOSCORE			0.32**

Note: ** Significance ≤ 0.01 (2-tailed)

3.7   DISCUSSION OF RESEARCH DESIGN

This study has two basic objectives: the verification of current insights ('conjectures') on the basis of hypotheses and the development of a general (non-sector-specific) theoretical model. For this reason, we have chosen a matched sample model. Considering the redundancy which is so necessary for theory-forming research, a method has been set up which makes it possible to measure a large number of variables. A matched pair study set up in this manner has well-known advantages: it leads to general conclusions; many other explanations can be eliminated via matching; and it requires a relatively small sample. The charm of matched pairs research is evident in the research reports. It is attractive to readers and is one of the much-travelled paths in international innovation research. Readers encounter 'hard numbers' and clear graphics which almost speak for themselves. By looking for identical twins with the same starting position, a whole maze of side-roads are cut off, leaving only the main traffic structure. Through the elimination (via matching) of situationally determined variance, perspective on general and non-company- or sector-specific relationships and insights can be acquired. It is thus possible to present concise and clear results which display a high degree of generalisation.

In the initial phase of the study, we put a lot of effort into following that model as precisely as possible. In practice, however, it appeared to be extremely difficult to find sufficient numbers of pairs of which both companies were willing to participate in the study. For practical reasons, then, we changed to a matched sample design. While this approach is not such a great problem for theory development, which has more to with building constructs than drawing hard conclusions, it does carry more risks with it where the verificative section of our study is concerned (the first objective). Continuous account has been taken of this fact in the analysis. In retrospect, these risks seem to have been acceptable: most results correspond to indications in the current literature. Moreover, the large number of variables enables the researcher to look subsequently for alternative explanations.

In addition, attention was devoted to triangulation in the analysis (Box 3.4). By questioning different persons in the company individually, the information obtained on the company will be more correct and precise. Moreover, a validation of the answers has been incorporated in the questionnaires by posing questions in both open and closed form. The success criterion for the dependent variable is also, as discussed in the previous section, checked for correctness and precision by including the judgements of the directors and researchers. Furthermore, since we employed a combination of various sources (internal and external to the company) we were able to triangulate the findings. This was done by checking for inconsistencies in the data and asking for clarification if inconsistencies were found. We did not, however, follow a formal triangulation process, but instead relied on group problem-solving sessions which checked whether any cases contradicted the findings. In addition, we explicitly chose to cross-validate the findings in follow-up studies (of which two are already under way). There is thus no reason to assume that a systematic bias has occurred. Finally, a more verificative study will be carried out in subsequent projects. One might refer to this as the 'external triangulation' (cross-validation), since it takes place outside the framework of the research project (although within the framework of the research programme). The important thing is that more attention can be devoted to processes over time (longitudinal) in this way. For that is the greatest limitation in this approach: to a significant degree, it is a glimpse of the situation at a certain moment in time.

BOX 3.4 TRIANGULATION

Triangulation is a method that originated in sailing, where it is used to establish the precise position of a ship (Hertog and Sluijs, 1995, p. 247). For someone wanting to locate their position on a map, a single landmark can only tell them that they are situated somewhere along a line in a particular direction from that landmark. With two landmarks, however, the exact position can be pinpointed by taking bearings on both landmarks; they will be at the point where the two lines cross. In social research, reliance on a single piece of data brings with it the risk that undetected error in the data-production process may render the analysis incorrect. If, on the other hand, diverse types of data or a combination of methodologies in the study of the same phenomenon lead to the same conclusion, one can be a little more confident about that conclusion. Such confidence will be well-founded to the degree that the different kinds of data have different types of error built into them. Thus, by making use of different case studies, more than one informant, a combination of external and internal informants or different data-collection techniques, one can increase the reliability and utility of the results of the study. Triangulation serves in particular to validate the results and to bring depth to the description of the phenomenon (see Hammersley and Atkinson, 1983).

From a methodological perspective, we can conclude that the most ideal methodology would be to gather longitudinal qualitative and quantitative data on random samples. However, given our objectives (general critical success factors of innovative companies applicable to a broad range of industry and service sectors), this remains very difficult to accomplish in practice. This kind of intensive research with large samples runs into practical financial and personnel problems. In addition, it can be expected that such a design will be confronted with high non-response rates. Therefore, a new practical compromise has been sought: the matched sample case survey. As has been discussed, this is not the beaten path for this kind of research.

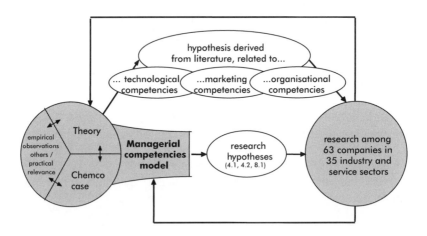

Figure 3.6　Schematic representation of our research process

Our expedition in the swamp had an iterative quality (see Figure 3.6). We went back and forth between current theoretical concepts, observations by others and our own observations and ideas. After a discussion of the empirical observations by others, the practical relevance and current theoretical concepts of Chapters 1 and 2, Chapter 4 will present an exploratory case study (the Chemco case) which forms the basis for our theoretical framework: the managerial competencies model. Two research hypotheses emerge from this framework (4.1 and 4.2) and a third research hypothesis is added in Chapter 8 (research hypothesis 8.1). In Chapters 5–8 inclusive, current theory related to each type of competency will be discussed. Based on existing research, a number of hypotheses will be formulated and validated for each chapter. Validation will be focused on determining whether these (usually single-sector-based) hypotheses also hold for a sample with companies from 35 industry and service sectors. The findings will then be correlated with the literature and the competencies model.

It is important to recognise that there are two types of hypotheses:

- hypotheses based on insights from existing literature (critical success factors). These insights are tested for their non-sector-specific validity;
- hypotheses we have formulated ourselves on the basis of the exploratory case study (4.1 and 4.2) and our findings during the present study (8.1). These go further than the current knowledge.

NOTES

1. The research project entitled 'Innovation management in the process industry', as reported in (among others): Cobbenhagen et al., 1990a, 1990b; Hertog et al., 1996.
2. Triangulation is defined as the combination of methodologies in the study of the same phenomenon. It serves in particular to validate results (Hammersley and Atkinson, 1983). The term 'triangulation' is derived from a loose analogy with navigation. To locate one's position on a map, a single landmark can only indicate that one is situated somewhere along a line in a particular direction from that landmark. With two landmarks, however, the exact position can be pinpointed by taking bearings on both landmarks; one is at the point where the two lines cross. In social research, reliance on a single piece of data brings with it the risk that undetected error in the data-production process may render the analysis incorrect. If, on the other hand, diverse kinds of data lead to the same conclusion, one can be more confident with that conclusion. Such confidence will be well-founded to the degree that the different kinds of data have different types of error built into them.
3. Scientists do not attempt to select 'representative' experiments either.
4. See subsection 3.4.2 (matching the pairs) for the selection procedure of the pairs by the experts.
5. In this connection, a comparison can be drawn with the so-called expert systems in decision theory.
6. In one case, in which the number of pairs to be selected was greater than the number of sectors in the class, two pairs in the same sector were selected.
7. The reasoning behind this is that innovation management success factors are, to a certain degree, dependent on the size of the company. In particular, there is a difference between the small companies (up to 50 employees) on the one hand and medium-sized and large firms on the other.
8. The questionnaire (in Dutch) is available from the author upon request.
9. Stevens (1946, 1951) proposed that measurements fall into four major classes: nominal, ordinal, interval and rational, and that these levels allow progressively more sophisticated quantitative procedures to be performed on the measurements.
10. A regression model will be presented in Chapter 8. On conclusion of the study, all companies which declined to participate were sent an abridged questionnaire containing the questions which enabled us to determine the variables for those companies used in this regression model. Subsequently, it was established which of these companies would be

considered as front-runners and which as pack members on the basis of this model. Eleven companies were involved here which did not participate in the main study but were indicated by the experts. These firms indicated that they would only participate if the research involved a postal questionnaire which could be answered in a reasonably short time. At the onset of the research we did not have this possibility, but after the results of the main research project were analysed and before anything had been published about them, an abridged questionnaire consisting of the most significantly differentiating questions was prepared and sent to the firms which initially declined to participate. It was returned by 11 of them, bringing the total response from firms to 74 (which is a total response rate of 82 per cent) and the number of complete pairs to 31. The data from these 11 companies supported the validation of the external criterion as well as the other results to a significant degree.

4. Sketching the model

> My personal opinion is that our company culture is not the right one to work with small scale production facilities, with small customers and with small subsidiary companies. We are too bureaucratic for that, we have too many layers and too many rules. (A. Olivier, General Manager of Shell Nederland Chemie, explaining Shell's motivation for terminating its investments in specialty chemicals; in *De Volkskrant*)

4.1 AN EXPLORATORY STUDY

4.1.1 Objectives

In Chapter 2 we sketched a framework of capabilities and competencies. A distinction was made between operational capabilities, key capabilities, managerial competencies and core competencies. It was indicated that our research project would focus on the managerial competencies. In this chapter, we shall elucidate these managerial competencies more clearly through an intensive exploratory case study conducted prior to the main study. This exploratory study had three objectives. Besides providing us with building blocks for managerial competencies and a tentative framework explaining the differences in innovative success which links these building blocks, it was also intended to provide us with insights into the dynamics of competencies. These aims will now be discussed in more detail.

As discussed in Chapter 2, the current study is focused on the managerial competencies of firms, with the pilot case being used to operationalise the concept of managerial competencies. The first objective of the exploratory study was thus to refine and operationalise the concept of managerial competencies into subconcepts such as processes, critical success factors and capabilities in order to help us study this phenomenon. Whereas most studies on (core) competencies (such as Prahalad and Hamel, 1990) focus on the technological aspects of competencies, innovation literature teaches us that the organisational and managerial aspects of innovation should not be undervalued in explaining differences in innovative success. The case should therefore highlight the importance of managerial and organisational aspects.

In order to learn more about the importance of these aspects as opposed to technology-related aspects, we decided that the initial case study should be of a company from a technology-intensive industrial sector, the chemical industry. This would allow us to illustrate the importance of organisational and managerial competencies in a technology-intensive and rational environment. So the first aim was to provide ourselves with building blocks for managerial competencies which may play major roles in explaining differences in innovative success.

Second, and closely linked to the first objective, these building blocks for managerial competencies would then be used to build a tentative framework of organisational and managerial factors explaining differences in innovative success. This framework would then serve as the starting point for our main study. Based on this case study, we would be able to make the concept of managerial competency into a framework within which a wide range of (usually fragmented) theories can be placed. This implies categorising potentially useful concepts and indicating their linkages (coding). Besides providing an integrating framework, the case study was also intended to generate new ideas and verify the appropriateness of the problem formulation. The starting point in this respect was to find ways to translate the concept of managerial competency into practical situations. Given the overall goal of the study, it was imperative for these concepts to be stated in a language which is pertinent to the day-to-day world in which managers operate, so that they will understand the questions and recognise the answers. The important factor in this connection was to keep both feet on the ground and observe what took place in practice.

Third, the exploratory study was to provide us with insights into the dynamics of competencies. We wanted to learn more about the initiation of competencies and the contexts in which they could develop. To that extent we had to find a fairly complicated (messy) case in which all possible factors could play a role and, in particular, in which the company development could be clearly traced and showed striking changes over time; short, an extreme case with a redundancy of events and factors.

4.1.2 Selection of the case

Considering the vast range of manufacturing and service sectors to be studied in the main study, the exploratory study had to involve an innovative company which displayed a great diversity of facets related to innovation management: facets related to products, processes, services and organisational innovations. It did not necessarily have to be a successfully innovating company. Preference was given to a large, multilevel, multiproduct company. Since such companies are usually complex, we would have the opportunity to study a great variety of aspects from different angles. Furthermore, we were looking for a company with a rich history of innovative attempts so that we could study its innovative endeavours in retrospect.

In order to select a company which met these criteria, it was necessary to have an idea of the main lines of the company's story in advance. For this reason, we decided to choose a company with which we had already become acquainted in the course of previous research. We found such an example among the 11 European firms studied (in four countries) as part of the research project on 'Management of innovation in the processing industry', for the execution of which the author had been partly responsible. In this project, we had obtained a reasonable picture of ten business units from diverse companies in the European processing industry. Based on the above criteria, we ultimately selected the 'fine chemicals and specialty products' business unit (FCSP) of one of these companies as the potential case. This case, with which, after four interviews, we were already reasonably familiar, could then be investigated more extensively in terms of both depth and breadth by means of a large number of new interviews, documentary analysis and feedback sessions. The management agreed to permit further investigation into the company on condition that the company's name would not be mentioned in publications. Thus, for reasons of anonymity, the real name of the firm has been changed to Chemco. We have also changed certain company characteristics which were not relevant for the analysis but which could reveal the firm's true identity.

Chemco is a large European-based chemical firm with a strong position in commodity chemicals. According to industry experts, it has outstanding technological competencies in many commodity chemicals. With the aim of diversifying into fine chemicals and specialties, a group (later to be called a 'business unit': BU) was formed in the late 1960s with the goal of further processing certain side and waste products of the commodity plants into intermediate products with a higher added value. The group was given the name: Fine Chemicals and Special Products Group (FCSP). The BU, originally launched to upgrade side products, developed after some 25 years into a medium-sized intermediate producer of fine chemicals.

At the time of the field study (January 1992), the Fine Chemicals and Special Products unit had reached a consolidation phase, which allowed for reflections on the recent history of the firm. Chemco provided us with the opportunity to study its innovative efforts over the past 20 years. The focus was to be on the organisational and managerial aspects of innovation and growth. This company was selected particularly because, like many of its competitors, it had gone through a significant transition in innovation focus during the two decades prior to the study. This transition had to do with the strategic aim of becoming less dependent on the cyclically sensitive commodity products and focusing more on specialty products. It implied a dynamic change in the direction of product innovation within a more complex business environment than the company had ever faced before. The case thus includes many aspects of the process of managing innovations. From an organisational perspective, it is intriguing to study the problems that large commodity-producing firms face in cultivating a viable fine chemicals or specialties division. However, the goal of this exploratory study was defined more broadly than a study of organisational difficulties faced by a commodity company in diversifying into specialties. It was to

> Design a theoretical framework which can be of help to managers in creating successfully innovating organisations and which can become the basis of a theory explaining innovative success.

4.1.3   Method of exploratory study

As the aim of this exploratory study was to construct theory, we decided to adopt the case-study methodology. Case studies are generally regarded as an adequate type of methodology for building theory (Hertog and Sluijs, 1995, pp. 134–140). As was discussed in Chapter 3, Yin (1984), for instance, states that case studies are the preferred (research) strategy when 'how' or 'why' questions are being posed, when the investigator has little control over events, and when the focus is on a contemporary phenomenon within some real-life context. Case studies enable the researcher to gain a good understanding of the process (the 'how' question: the story) and the context. He (Yin, 1990, p. 23) defines a case study as: 'an empirical inquiry that investigates a contemporary phenomenon within its real-life context when the boundaries between phenomenon and context are not clearly evident and multiple sources of evidence are used'. Case studies are especially helpful in situations in which knowledge about the phenomena studied is fairly slight or where the phenomena are so complex or elusive that the relevant variables cannot yet be clearly identified (Hoesel, 1985; Hertog and Sluijs, 1995). One then makes use of empirical observations to construct a theory via induction. As case studies are mostly concerned with change processes whereby the phenomena which appear today have their roots in the past and in turn influence the future, Hertog and Sluijs (1995, p. 134) argue that case studies should also be concerned with: 'the placement of a phenomenon in an historical perspective'. In other words, case studies are not only concerned with the description of a situation at a given moment in time, but also with the analysis of the process which led to that moment (Weiss and Rein, 1972, cited in Hertog and Sluijs, 1995, p. 134). Thus, the objective is to write down the story (Box 4.1).

BOX 4.1 KNOWLEDGE IS ONE THING, CHANGING THINGS ANOTHER

During the intake interview, the current business unit manager suddenly said: 'the funny thing is that we have a very good idea of the things we have been doing wrong in past decades or are still doing wrong. These things have been stated in various documents, some even dating back to the 1970s. But, and actually this is the tragedy of it, we never seem to be able to change things the way we tell ourselves they should be changed'. At that moment, we were convinced we had selected the right kind of case, because it was not knowledge of the ideal situation that this firms was lacking but knowledge about the way to get there. And this implies knowledge about the complete picture which makes one organisation successfully innovative and the other mediocre. These were the aspects we wanted to document, to gain insight into 'the whole story'.

The theory (or theoretical framework) we intend to construct will be built on the empirical observations. The methodological process by means of which this can be achieved is described in the work on grounded theory building (Glaser and Strauss, 1967; Strauss and Corbin, 1990). The grounded theory approach prescribes a comparative method of constructing theory from case study research. The method relies on continuous comparison of data and theory: 'it is discovered, developed, and provisionally verified through systematic data collection and analysis of data pertaining to that phenomenon' (Strauss and Corbin, 1990, p. 23). The basic principle of the grounded theory approach (Hertog and Sluijs, 1995, p. 136) is that new theories in the social sciences are much more likely to be developed on the basis of observations of reality than through abstract thinking behind a desk. Glaser and Strauss (1967) see theory development as a process in which new concepts are created from observations. From this perspective, relationships between concepts and phenomena develop out of a sort of balancing act between theory and observation. During the development of a theory, the theory maintains a double significance. At one moment it is an interpretation of reality, at another it is an explanation of that reality. The final product is an explanatory theory which can be verified by research (Hertog and Sluijs, 1995). The method emphasises both the emergence of theoretical categories solely from evidence and an incremental approach to case selection and data gathering (Eisenhardt, 1989). As our intention was only to identify concepts that might explain innovative success (in order to build a tentative framework), there was no need to follow the grounded theory approach to the letter. We therefore relied on a slimmed version.

In line with this methodology, our aim was to describe the process FCSP went through in the past decades as concretely as possible. Subsequently, remarkable changes and phenomena would be interpreted. Strauss and Corbin (1990, p. 143) define a process as 'the linking of sequences of action/interaction as they pertain to the management of, control over, or response to, a phenomenon'. In this view, the actual events are placed on a time scale, so that a clear story line (a chain of causes and effects) emerges (Hertog and Sluijs, 1995, p. 133).

Interviewees were selected among people who had worked for FCSP during the period under investigation. Selection was done by the current management team of the business unit in close co-operation with the researchers. The respondents can be regarded as a good representation of the various functional department and management layers of FCSP during the 1970–92 period. Many of them no longer work for this unit and have had time to reflect on the past situation. In addition, both the former BU manager and the former R&D manager of a sister business unit from the same division as FCSP were interviewed, as was the director of Chemco's Corporate Planning department and the director of Chemco's Corporate R&D department. In total, 17 people were interviewed (Appendix 3), some more than once.

The interviewees were asked to tell their story. No fixed questionnaire was used. Questions were posed for clarification or when inconsistencies with respect to other stories or data arose. Each interview took between two and three hours. A few interviewees were interviewed twice. Secondary sources (internal documents, internal studies, business periodicals, press releases and external consultancy reports) were studied to obtain additional information and validate the findings.

As the interviews were undertaken over a time span of about three months, the researchers were able to test their findings in an iterative manner, so that a framework of success factors gradually emerged.

After each interview, passages from the interview report were coded according to the type of strengths or weaknesses they revealed with regard to innovation management as well as internal or external facilitators or blockers. All interviews were conducted by the same pair of researchers. The coding was done by each researcher independently of the other. These individual coding schemes served as the basis for the final coding scheme drawn up on the basis of consensus between both researchers and compared to recent literature on innovative success. The results of this coding process are discussed in section 4.3. The final grouping codes formed the basis for the questionnaire for the main study of this book. In the light of the main themes to be deduced from current literature and the specific focus of the study, a few codes were eventually dropped as inputs to the questionnaire (for example, finance).

The results of the analysis were tested in several feedback sessions. The findings were first discussed with the business unit manager and the corporate R&D manager (in separate meetings), after which they were presented during a workshop to which all interviewees were invited; 12 actually participated. The purpose of this workshop was dual: first, it served to provide formal feedback of the results to the participants; second, it was used to test some of the outcomes. As such, the workshop was part of the coding process.

4.1.4 The chemical industry

Before beginning with the case description, we shall first devote some attention to the structure of the changes in the chemical industry in general. After all, this is the context within which the developments described in the following pages take place. The reader is thus provided with the framework within which the changes that the firm has gone through can be placed.

The chemical industry is a science and technology driven industry (Pavitt, 1984). The large players are technology-based firms that have been dedicated to science, technology and innovations for decades (Cooper and Kleinschmidt, 1993a). Many different sciences are involved; new chemical products are often based on inventions and not on product (re)design or engineering, as is the case in many other industries. Over the past decades, increasing ecological requirements, a globalisation of the industry and growing competition have been boosting (the need for) innovation in both products and processes even more. Thus the need emerged for tools and insights to manage innovation processes effectively (Cobbenhagen, et al., 1990a; Cobbenhagen and Philips, 1994). In fact, the ability to develop dedicated products at a fast pace has become an important strategic weapon for firms in the chemical industry.

From a management perspective, batch volume and added value are important categorisation criteria in the chemical industry. Based on these criteria, the chemical industry (anno 1992) can be (roughly) divided into commodities, fine chemicals and specialties (see also Table 4.1). The differences between these subsectors reveal themselves in the amount of strategic attention devoted to innovation in products, processes and technology and the extent to which they are market-driven.

Innovations in the mature and stable environment of the *commodities* subsector are orientated towards the improvement of processes and the search for new applications for existing products rather than the development of new products (and processes).

Specialties, in the strictest sense, are chemical products sold on the basis of their performance rather than their composition. They are often 'tailor-made'. Opportunities in these markets thus depend on one's capacity to respond to the dynamics of innovations and technological developments. Being able to innovate quickly is very important in this subsector, since the product life-cycle for individual specialty chemicals is quite short. A low time-to-market is essential for this type of product. After all, the customers are pressed for time in their own production processes.

Fine chemicals are used as 'pathways' or intermediates to other products and are often the raw materials for specialties. Fine chemicals, like commodities, are sold on the basis of their composition and are generally interchangeable with other products of the same composition. Characteristics of the fine chemicals industry are a continuous flow of new products and a relatively high price/volume ratio. R&D is focused on both process and product development and improvement. Competition is often related to price, but image and service are important distinguishing features as well. Although fine chemicals and specialties each have their different definitions, in practice the line between them is quite vague.

In the 1970s, a change of strategic focus became apparent. In the attempt to avoid the intense competition on the commodity markets and to become less sensitive to cyclical changes, the strategic focus of many chemical producers shifted from commodities towards customer-orientated, more refined chemicals with a higher added value: fine chemicals and specialties. As a result of the large variety of products and customers, these sectors are less cyclical and usually offer greater stability and profitability (Table 4.1).

The shifting strategic (and market) focus also led to a different focus on innovation, moving from process developments and process innovations towards product innovations. In the process, marketing became a new and influential player in the internal innovation game. For specialty and fine chemical products, adaptation to and focusing on consumer demands (product modifications) is very important: much more so than for commodity chemicals, for which a high 'on stream' is important; so the focus of innovation in this part of the industry has therefore been on improving the production facilities, hence process innovations. There are virtually no differences in the same kind of commodity product from two different producers. It was thus virtually impossible for a firm to distinguish itself from its competitors on the market.

Table 4.1 Main characteristics of the chemical industries

Characteristics	Commodities	Fine chemicals	Specialties
General Characteristics			
Life-cycle of products	long	moderate	short/moderate
Number of products	x00	x.000	x0.000
Product volume	> 10.000 ton/year	< 15.000 ton/year	highly variable
Product price	< US$5/kg	> US$5/kg	> US$5/kg
Product differences	none	very low	high
Added value	low	high	high
Capital intensity	high	moderate	low-moderate
R&D aimed at	process engineering	process engineering and product engineer	product engineering
Success factors (relative importance):			
Cost position	high	average	low
Technical service	—	average	high
Close contact with customer	—	high	high
Examples of products	fertilizers ethylene benzene	engineering plastics pharma intermediates epoxys	flavours and fragrances food additives electronic chemicals

Source: Based on Theeuwen and Polastro (1989)

One area where a competitive edge could be achieved was production: producing (more) efficiently against low(er) costs, minimising shutdowns, and so on. This environment calls for an entirely different style of organisation, in terms of both culture and structure, than an environment in which the focus is primarily on the market and where competitive advantage is primarily to be found in adapting products to meet the wishes of customers. Furthermore, specialties and fine chemicals are much more service-orientated and customer-orientated than commodities. The 'product' is more than the chemical substance. Firms tend to regard their products as 'offering solutions to customer problems'.

In fact, commodity firms which diversified into fine chemicals and specialties faced the challenge of managing two completely different cultures in one company (Winsemius, 1988): the commodities culture, which, like American football, is hierarchical and tightly controlled, and the fine chemicals culture, which is more like basketball, improvising, flat and flexible. Consequently, fine chemicals and specialties needed to be managed and organised differently than commodities. However, these types of organisational structures were new to most large chemical firms, whose routines had been dominated by commodities.

What has become clear from various studies (Anderson and Ayers, 1992; Cobbenhagen and Philips, 1994) is that there is a great difference in performance with respect to specialty and fine chemicals activities between, on the one hand, firms which originated in these markets (or have a strong position in more customer-orientated products, such as Henkel) and, on the other hand, firms which were originally commodities producers and later diversified into specialties and fine chemicals. The large commodity-dominated firms appear to have been less successful in fostering specialties and fine chemicals than the (often smaller) firms where specialty or fine chemicals operations predominate. When moving towards specialties and fine chemicals (whether through acquisition or internal growth), companies sometimes faced completely new technologies and markets. Some of the large chemical firms were unprepared for the new technologies and requirements that the new markets imposed. The danger of diversifying into these unknown territories has been extensively documented in the innovation literature (Anderson and Ayers, 1992; Roberts and Berry, 1988).

Since the late 1970s, when many commodity producers diversified into specialties and fine chemicals, many things have changed. For one thing, these 'new' sectors have become more competitive and less profitable than they used to be. This is partly a result of the mass movement of commodity companies into these areas and partly because many of these markets are no longer considered special. Many companies have been able to master the required technological and marketing competencies. Another factor is market saturation and maturity. While the answer to some of these growth problems is new products for new applications or new markets, many chemical companies have had great difficulty in moving outside their original area of expertise (Warthouse and Louie, 1989). The comfortable situation of most chemical firms, where anything R&D could invent and production could produce was sold on the market, had ceased to exist. Coping with this new situation means devoting attention to the organisational and managerial aspects of innovations. However, chemical firms had been so successful in the post-war period with their robust, functional R&D organisations that they often failed to see the need to change their routines (Cobbenhagen and Philips, 1994).

4.2 CASE DESCRIPTION OF CHEMCO'S FCSP BUSINESS UNIT

The history of Chemco's FCSP business unit can be roughly divided into four phases: start-up (1996–72); pioneering (1972–76); expansion (1976–88) and consolidation (1989–92).

4.2.1 1969–72: start-up

Chemco, which until 1969 had mainly produced commodities and basic feedstock, was looking for a way to upgrade the by-products of some of its commodities. According to a 1969 internal document, corporate management saw huge potential for small (in terms of global sales volume), customer-orientated products because of the high profit margins these products could generate; their technically simple production process; and the expected easy distribution through existing sales channels and internal use within Chemco. They were regarded as attractive extra business possibilities with minimal additional costs. The question was whether these small products were merely a side issue or whether they could become a cornerstone of the company. In 1969, therefore, the FCSP (Fine Chemicals and Special Products) group was formed and assigned the task of developing products which could be produced and sold with limited research or research closely linked to existing studies; with limited investments and a short pay-out time; and with limited marketing efforts.

By 1972, it had become clear that a short pay-out time was too severe a requirement. Only very few products could meet it, and those that did had virtually no possibility of becoming cornerstones. Consequently, the Chemco board of directors decided to allow FCSP to develop activities requiring longer pay-out times and larger investments. In addition, the research budget was increased. This marked the beginning of the pioneering phase.

4.2.2 1972–76: pioneering phase

The FCSP group started with a very strong feedstock position, which was generally regarded as one of its most important strengths and allowed FCSP to create a modest position as a producer of 'me-too' intermediates in the early 1970s (intermediary chemicals that were similar to products already marketed by competitors). The pioneering years led to fast growth, but it was not as fast as hoped for. In retrospect, it was concluded that the turnover objectives were not always that realistic. In the late 1970s, it became clear that the ambitious growth objectives would never be achieved by internal growth alone. One reason for this was that the available marketing and research potentials were insufficiently focused on performance chemicals (which had become one of FCSP's new areas of attention in addition to specification chemicals). So the market was scanned for possible takeover candidates.

However, FCSP lacked a clear strategic focus. For one thing, this was a result of the strong feedstock position which hindered the development of a clear strategic market focus that could guide the generation of new ideas. From the beginning, their mission had been to create new products on the basis of the available feedstock provided by the Chemco commodity divisions. This resource push strategy prevented the development of a market pull strategy. New ideas were based much too heavily on the feedstock and much too little on the markets: 'Until 1975 we had no idea what our markets were. We knew our pots and pans and we knew what our feedstock was. We couldn't care less where we left the stuff' (former FCSP marketing researcher).

Looking back on the pioneering period, interviewees tended to describe the group as 'cowboys exploring new territory'. The staff were generally quite young and very enthusiastic. They had a mission: 'To prove that experiment will lead to new specialty products' (internal document, 1990). In this phase, FCSP had a simple structure with a single-headed leadership. The unit operated quite independently of the existing Chemco activities. Its way of operating can be characterised as flexible and informal. Improvisation and informal mutual adjustments were much more important than formal (Chemco) systems. In 1975, the Chemco company adopted a divisional structure and the FCSP group became a business unit within the Industrial Chemicals division, losing its separate status. Its new place in the hierarchy can be regarded as resulting from a lack of understanding of the specific characteristics of this business and a lack of appreciation of its particular needs. The wild years were over. The Chemco top management concluded that FCSP had become adult and had to be incorporated into the Chemco organisational structure.

4.2.3   1976–88: expansion phase

In the late 1970s, partially due to the oil crises and the generally poor economic structure of the bulk chemicals industry, objections were put forward about the concern's heavy growth in commodities. One of the characteristics of bulk chemicals is that the raw materials account for a great part of the sales price, so there is a heavy dependency on the raw materials (primarily naphtha). The oil crises thus dealt the chemical industry a double blow: in the energy bill and in the feedstock costs. There was increasing awareness of the fact that the concern had to diversify into products for which the raw materials formed only a small part of the sales price. The corporate management decided to speak out explicitly in favour of further development of knowledge-intensive products, the specialties and fine chemicals. In future, according to the goal set by management, about one-third of Chemco's turnover would come from specialties and fine chemicals. Within this new Chemco strategy, the FCSP unit gradually gained a more important role. The focus was now on growth with new products. The important task which FCSP set for itself for the 1976–80 period was to establish a sound and sufficiently broad base for the reinforcement of Chemco's position in the field of special and fine chemicals after 1980 (centre of growth function). In a 1978 internal report, management indicated that the following conditions would have to be met in order to achieve this goal:

- more attention would have to be devoted to building up the necessary technical know-how in R&D and marketing;
- greater flexibility would have to be effected in the production installations, for example by constructing a multipurpose plant in advance;
- balanced growth of the organisation would have to be ensured, whereby the attraction of other Chemco units should be resisted as effectively as possible so that built-up expertise would not leak away via too intensive job rotation;
- there would have to be sufficient financial leeway to make it possible for a vision of the future to prevail over short-term profits.

But the belief in the possibility of achieving expansive and profitable growth in these relatively unknown market segments solely on the basis of internal developments was openly questioned. Therefore, a small company with a complementary product range was acquired in 1978.

Within Chemco, FCSP had created a kind of maverick position: self-willed professionals who creatively explored the technological boundaries with little respect for the Chemco way of operating. Over time, the FCSP organisational structure evolved towards a functional structure. With this change in structure came new rules and regulations. The people who were developing, producing and selling new products became organisationally (and later physically) separated from the people producing and selling existing products. Tasks, authority and responsibilities would be defined more clearly. The culture changed gradually as well, since the new structure left hardly any room for the old cowboy mentality. But tasks, responsibilities and authority remained vague, mainly because of the rapid growth. And despite the intended decentralisation of authority, most decisions were still taken by unit management.

In 1982 it became clear that FCSP was not on the right track (Table 4.2). Problems arose particularly in connection with the technology, the market and the organisation.

Table 4.2 Main findings from an internal SWOT analysis conducted in 1982[a]

	Strengths	*Weaknesses*
Technology	• strong position in a number of technologies	• technologies were all isolated
	• sufficient differentiating technologies	• lack of the most common multipurpose equipment
	• R&D capacity	• little flexibility
Marketing	• risk-spreading through large number of markets	• shallow knowledge due to large number of markets
		• insufficient market orientation (no clear choice of segment)
Organisation	• versatile personnel	• tasks and authorities unclear
	• young organisation which is loyal and motivated	• *ad hoc* operation/poor setting of priorities
	• improvisation	• staff turnover too high
Strategy		• no vision
		• changing strategy
Finance	• strong financial capacity of mother company	• no insight into project costs

Note: [a] Categorised by the researchers.

Technology

For one thing, FCSP lacked the flexible hardware which its competitors had. This prevented effective and efficient scaling up of new products. In the fine chemicals and specialties industries, the possession of multipurpose installations is vital for success, even if this implies some overcapacity. FCSP lacked such flexible plant, which had no place in a bulk culture. This was a definite bottleneck in the early years. It was not until the late 1980s that the FCSP management felt it had sufficient installations to complete its mission. The problem which then had to be faced was that the organisation was 'not ready' to deal with the new technology, which was developing faster than the organisation and the marketing. Due to the great amount of hardware simultaneously coming on stream and a number of setbacks in some markets, FCSP was forced to respond to as many market demands as possible in order to ensure that the installations would operate rapidly at a profit. The result was a zigzag movement in the markets. Profitable production succeeded only partially.

Marketing

For a long time, there was a lack of focus in the market. The unit tried to respond to virtually every market demand, developing a whole spectrum of activities in the process: 'The simple fact that there was a potential customer was enough for us to serve him' (former marketing manager). However, there was little insight into the costs of all these products. Furthermore, FCSP did not have enough knowledge about the end markets. Knowledge about the size of the (potential) market and the position of competitors was very scanty. The formation of a market network was one of the new strategic issues. While speed of delivery was important in these markets, FCSP was too slow in responding to a question with a sample product:

> This is partly the fault of the companies we were supplying. They were very close-lipped and provided, for example, no insight into the [end] product for which our product was necessary. But to make that first kilo, you need info on the end market. We were too slow. There were regular communication problems between Research and Market Research. For example, Market Research would often promise a sample, and only then bother to find out if the product could be developed. (Former researcher)

By the late 1980s, it was clear that things could not go on like this and more effective screening processes began to be employed. Technological spearheads were formulated, based on the unit's technological strengths.

Organisation

There was a growing awareness of the need for organisational and strategic change. During this phase, some organisational adaptations were made without deviating too much from the commodities organisation. In order to achieve the ambitious growth goals of FCSP by means of internal developments, the New Developments department was formed in 1985. Within this department, several product managers were appointed. It was intended that new development would be organised project-wise. This, however, was not the case:

> Project managers were appointed and it seemed as if we would work project-wise. I stress the word 'seemed', because we did not work project-wise at all. We simply had no project organisation. Project leaders were not given clear authority and could not make binding commitments. Almost everything was done on a voluntary basis... or simply not done at all. (Former product manager)

In 1987, the product managers declared that the project-wise manner of working was not a success. Agreements were not being kept. Their tasks, responsibilities and authority were clear neither to themselves nor to the rest of the organisation. And because the project leaders lacked a formal budget, they had no insight into the costs and man-hours spent on projects. Despite all the good intentions that led to the formation of the New Developments department, the functional managers continued to call the shots. Project leaders had to lobby for co-operation from other functional departments. Consequently, projects proceeded in a very *ad hoc* manner. According to some people, this was an unavoidable intermediate step towards a more professional organisation. A step, however, which lasted much too long:

> FCSP developed out of a gang of adventurers, and we had to gradually grow into a normal (that is, bulk goods) organisation. But we realised too late that the organisation had to change. (Former product manager)

Furthermore, the influence of the product manager was restricted to the early phases of the innovation process. When the product was transferred to production a new project manager was appointed, whose way of doing things did not always correspond with his predecessor's. Project managers were judged very harshly if they exceeded their budgets. Consequently, they usually made cuts in provisions for future facilities (such as the vital multi-purpose installations). The New Developments department seldom provided project managers with feedback in this respect.

Another problem at FCSP was staff turnover, which was enormous during the first two decades, particularly in the marketing department and management. People who were successful at FCSP were rapidly snatched away by top management and placed in the larger divisions. At that time, the FCSP organisation was not really taken seriously. It had somewhat the status of a Chemco training site.

If we had to identify one core problem at the root of many of FCSP's problems, it might very well be the tension between specialties/fine chemicals and commodities. Apart from the pioneering phase, Chemco had always tried to keep FCSP harnessed to bulk production. In that period, FCSP was regarded as something unimportant, small-scale, a sort of game. The 'techies' had the upper hand, 'marketeers' were secondary. The organisational structure was functional, with an overabundance of bureaucratic procedures to be dealt with. The culture of the commodity divisions dominated the governance of FCSP again. In retrospect, the present manager of the business unit concludes that the management at that time did not realise that fine chemicals and specialties require a different approach from bulk production: different in terms of technology, production, client relations, management style and organisational structure, but also in terms of company culture. A specific specialties/fine chemicals culture, however, such as could be seen in competing companies which had originated as producers of specialties and fine chemicals, was never able to develop at FCSP.

4.2.4 1989–92: consolidation: streamlining and optimising production processes

The expansion phase, and especially the years between 1982 and 1988, was characterised by an expansive growth with the focus on the internal development of new products. During the whole expansion phase, the total number of products rose from 20 to 60. Looking back, one might state that too many new products were developed. The interviewees generally agreed that FCSP had reacted to too many new ideas without considering the consequences. At that time, however, this manner of operating was seen as a way to find one's own place in the market. All these new developments made heavy demands on the group's finances. R&D costs and investment in processes had become much too high and profits eroded drastically. FCSP went into the red. By 1989 it had become clear that FCSP had neglected the development and improvement of its processes for too long. As one of the respondents put it: 'The advantage we had with our feedstock position was lost because of our outdated and inefficient plants' (former plant manager). In principle, this statement is indisputable, but it reveals only part of the truth. Here again, we are hearing from a respondent whose career had developed in the bulk sectors of the company. He looked for the cause of the problems in technology, in inefficient production plants. But the lack of market focus and the inadequate organisation were no less important causes of FCSP's problems.

It became obvious that this downhill movement would not stop by itself and that something had to be done. That 'something' was a drastic strategic shift, a move away from innovation and towards consolidation. This strategic shift was facilitated because the commodities sectors were performing surprisingly well. In retrospect, most of the sombre predictions made in the early 1980s about the future of bulk chemicals turn out to have been mistaken. By the late 1980s, partially due to scarcity on the market, bulk chemicals were enjoying a golden age, and they continue to play an important role. On the one hand, they are a great cash generator, in fact the most important source of funds for new developments. On the other hand, bulk chemicals are important from a strategic point of view, as the main provider of feedstock for the specialties. These developments, combined with the below-par performance of the FCSP business unit, lay at the foundation of the strategic shift. In 1989, division and corporate management decided that FCSP had to shift its attention from developing new products towards optimising and increasing the profits of the current product mix. The FCSP objectives for the first half of the 1990s were defined as follows (SMP 1990–95, internal document, 1989):

- consolidation (completion of the current mission, maximum utilisation of new plants);
- optimisation of product mix (improvement of low-cost position);
- provision of support to the unit's current cornerstones and selective growth.

A new business unit manager was appointed and innovation efforts practically came to a halt. No new risky projects were launched and a number of ongoing ones were stopped. Fixed costs were reduced wherever possible. The R&D budget was cut by a third and the New Developments department was closed down. Many R&D activities were redirected towards process improvements. Focus changed from the new to the current. Product innovations were clearly assigned lower priority in favour of process improvements and optimisation of current activities. This reversal bore fruit. FCSP's financial figures went up again. Although FCSP of course benefited from this financially, it is still questionable whether the unit was really set on the right track. Within the organisation today, people wonder whether too much knowledge about product innovations has been allowed to leak away.

As was mentioned before, FCSP is not a case on its own. In many large chemical firms, the initial enthusiasm for obtaining rapid success in fine chemicals and specialties has cooled and ambitious plans have been modified in the direction of less-ambitious, more-realistic scenarios. One can still wonder, however, whether FCSP, given a different approach, a different organisation, more experienced people and more freedom with respect to the Chemco organisation, might have been able to achieve more of its initial ambitions.

4.3   ANALYSIS

4.3.1   Coding

After each interview, passages from the interview report were coded according to the type of innovation management strengths or weaknesses they revealed as well as internal or external facilitators or blockers (see also section 4.1.3). All interviews were conducted by the same pair of researchers. The coding was done independently by each of the two researchers who conducted the interviews. These 'individual' coding schemes served as the basis for the final coding scheme drawn up on the basis of mutual consensus between both researchers and compared to recent literature on innovative success. This process took place during the period in which we conducted the interviews. The case was discussed in formal and informal meetings and gradually a picture of the theoretical framework emerged. Initial ideas were tested in subsequent interviews and some interviewees were revisited to fill in the blank spots.

Initially, some 30 codes were derived from the data. Through various discussions between the two researchers and a third researcher who was not involved in the study, it became clear that the codes could be grouped into technological, marketing, organisational, strategic and financial categories as well as one more rather elusive category which could be labelled organisational learning (Table 4.3). The resulting six grouping codes (or categories) formed the basis for the questionnaire for the matched-pairs study, the main study of this book. In the following sections, the analysis of the case will be discussed according to the grouping codes.

4.3.2   Technological competencies

Technology and hardware
Partly due to its solid position in bulk chemicals, Chemco has always been very strong technologically (the translation of ideas into production). This strength, however, turned out to be disadvantageous for FCSP because fine chemicals are subject to different criteria from bulk chemicals. For example, much heavier plants than necessary were constructed. Or, as a plant manager put it: 'We bought a Mercedes when an Opel Kadett would have been satisfactory'.

Table 4.3 Resulting codes

Codes	Categories (grouping codes)
Hardware/installations	Technological competencies
Position on specific technologies	
R&D capacity and potentials	
Transfer rate (Research → Development → Production)	
Use of external knowledge	
Knowledge build-up	Marketing competencies
Market orientation	
Knowledge of markets	
Knowledge of core values	
Sensing capacity	
Relation to sales	
Quality assurance	Organisational competencies
Business unit organisational structure	
Culture	
Interfunctional relations	
Authority problems	
Project management	
Personnel policy	
Strategic orientation	Strategy
Timing in company policy	
Vision/mission	
Financial problems	Financial aspects
Investment criteria	
Project cost targets	
Cash flow management	
No learning from past experience	Organisational learning
Not able to do what is intended	
Evaluation of projects	
Unlearning	
Barriers to learning	

Moreover, the high on-stream (plant capacity utilisation) requirement conflicted with the fitness for use principle which is so important for fine chemicals, the production of which demands a different optimisation criterion than high on-stream performance. Authorisation for new construction, however, was only granted when it was certain that 80 per cent of the capacity would be utilised. This capacity requirement could seldom be met and ultimately became a hindrance to the construction of the necessary multipurpose or multiproduct plants, where overcapacity is normal. To obtain permission none the less for construction, excessively optimistic capacity forecasts were presented. In 1986, however, these diversionary tactics led to the technology itself becoming a hindrance to innovation. The plants which had been built on the basis of unreal cost estimates were by then experiencing major overcapacity, and this was no longer tolerated by the division. Innovative efforts became focused more on keeping the plants running at full capacity than on seeking a strategic market position. Technology gained the upper hand in the development of new products and marketing could do nothing but follow. In this manner, the plants were optimised but not the profits.

Technological position

FCSP's technological position was limited in terms of the number and types of technologies. It lacked the most common multipurpose equipment for the execution of unit operations such as mixing, stirring, distilling, centrifuging and drying. Such activities are referred to as 'enabling competencies' (Leonard-Barton, 1995) or 'necessary competencies' (Hertog and Huizenga, 1997, p. 47), that is, competencies which the firm itself must possess in order to be able to work effectively and create added value. In this respect, they had clearly fallen behind their closest competitors. Moreover, most of the technologies they had mastered were not unique: 'FCSP relied on technologies which a fine chemicals and specialty producer simply has to have in order to be a player' (plant manager).

While they did have knowledge of a couple of unique technologies ('differentiating technologies') in addition, the possible uses for these technologies were limited. Due to the lack of a number of essential competencies, moreover, it was not possible to make full use of the differentiating technologies. Remarkably enough, Chemco FCSP had neglected its known technologies for too long.

R&D capacity

When compared with the average R&D budgets of many commodity business units, FCSP's R&D budget as a percentage of its sales can be classified as quite substantial. Until the consolidation phase, about 20 highly skilled (university level) chemists and technicians at Chemco's central laboratory devoted their attention to FCSP.

4.3.3 Marketing competencies

Marketing knowledge
FCSP was a newcomer in the specialties and fine chemicals markets. Besides commodity companies which, like Chemco, were in the process of diversifying into fine chemicals, FCSP also encountered a number of strong market competitors which already had long histories in 'customer-fit' chemicals. On these markets, reliability of delivery, quality and compliance with specifications were very important. FCSP, however, was unknown in the market, had a low 'security of supply' reputation and thus had an enormous amount of catching up to do. Moreover, FCSP was operating in a different type of market than was usual for Chemco. A different market approach was therefore necessary, although this was not fully recognised in the initial phases. Thus, FCSP not only started far behind its competitors, it also had great difficulty in catching up. Long-term relations with clients were the rule rather than the exception in these markets. To create such relations, experienced marketeers were needed, and that was exactly what FCSP lacked. It rapidly became clear that building up knowledge in the Market Research department would be crucial for success. Partly due to the large staff turnover, the unit had only limited success in this. The FCSP Marketing department was considered as an apprenticeship for rising stars within Chemco, and marketeers stayed no more than two years or so in the department. They had to deal with buyers who had been working in the sector for 20 years. The build-up of market knowledge thus remained inadequate for a long time.

Market orientation
For a long time, FCSP's strategic positioning left much to be desired. The unit was not clear about which markets it should concentrate on and thus responded to market demands in a highly *ad hoc* manner (an undifferentiated market approach). When overcapacity in the plants became a problem, it was decided to focus on technology, and marketing could only follow.

Separation between the marketing and sales organisation
The departments for marketing (responsible for new products) and sales (responsible for current products) were separate worlds. Innovations were sometimes resisted by the sales organisation. Furthermore, marketing was strongly technology driven.

4.3.4 Organisational competencies

Structure
Externally, FCSP employed a market specialisation and internally a technological specialisation. Co-ordination between the two clusters proceeded in a far from optimum manner. The New Developments department was supposed to provide project managers and had been assigned to stimulate and realise new activities. This construction, however, was not an unmitigated success. Marketing, Sales, Production and R&D were highly functional departments which had difficulty in handing over tasks to the New Developments department. The product managers often had no authority to get things done. Tasks and responsibilities thus remained unclear: responsibilities were taken, not given. This expressed itself (among other ways) in very frequent co-ordination problems, both within and between departments. The company suffered from a 'passing-the-buck' culture:

> We launched a discussion about tasks, authority and responsibilities. What was our place in the organisation? In our view, our place in the organisation and our tasks should be derived from the strategy. But the strategy was not clear. (Former product manager)

Culture
FCSP suffered greatly from being part of a bulk company. Chemco's bulk-orientated culture and structure formed significant barriers to a successfully innovating FCSP organisation. FCSP had grown out of a gang of adventurers which had to operate in a bureaucracy. Chemco thinks in terms of commodities, not fine chemicals. As a result, problems were tackled in the wrong way.

4.3.5 Organisational learning

Learning from past experience
FCSP learned virtually nothing in a structural manner with respect to the management of innovation. Each project manager took care of business in his own way. The continual friction between the departments was not resolved and led time and time again to delays and co-ordination problems. Due to the high personnel turnover, FCSP had no adequate organisational memory and, as a result, learning experiences rapidly faded away. Given the FCSP's learning function within Chemco, it is to be expected that these learning experiences were translated into business elsewhere in the concern.

The most important lessons learned by the management team in retrospect (at the time of our study) were:

- to devote more attention to synergy;
- to dare to make (clear) choices;
- to set priorities;
- to dare to tackle the organisation.

Evaluation

Interim project evaluations were usually omitted. Projects were only stopped if no agreement could be reached with the customer. This could sometimes occur in a very late phase of the innovation process. Even retrospectively, evaluations seldom took place. If projects were evaluated, the organisational aspects were seldom involved. Organisational problems were not mentioned by name and were allowed to lie dormant:

> The breakthrough in my own thinking came only after I had left FCSP. I then saw what we had done wrong, and these mistakes were particularly in the areas of organisation and strategy. (Former marketing manager)

Barriers to change

According to the former product managers, during the expansive phase there was definite resistance to innovation from the departments responsible for current products, Sales and Production:

> When I was hired here, I had the idea that I would have to carry quite a load. But in reality I often had a millstone around my neck. There was resistance to carrying that load. This resistance did not come so much from individuals. The interaction was pleasant, but not the organisation. (Former product manager)

For Production in particular, an average of six new products per year was too much. This department had an aversion to the continual introduction of new products. In consequence, there was constant fighting about who was really responsible for business: those involved with current products or those involved with new products.

4.3.6 Strategy

The FCSP ship navigated for a long time without a clear course. Strategic goals alternated at a rapid pace and the unit was sometimes said to have a 'windshield wiper' policy, constantly changing direction, but gradually progressing none the less. In fact, there was virtually no strategic direction at all. Most efforts were focused on the short term:

We had no carefully considered strategy. Around the end of 1982, for example, it seemed as though the strategy was to examine and acquire as many new products as possible. Requests which came in via Market Research were honoured whenever possible. Our task was seen as diversification, the process of raking in turnover. Little attention was given to synergy. Subsequently, this is precisely what people reproached us for. (Former R&D scientist)

The communication of strategy and the manner in which it was meant to be achieved left much to be desired. The lack of a clear objective expressed itself in competency fights and bickering. In retrospect, one can say that FCSP's initial assignment could definitely have been better specified, for example by combining the growth objective with profitability requirements.

4.3.7 Financial aspects

For a long time, FCSP was self-supporting, with very minimal profits. The division accepted FCSP's modest yields as long as the turnover continued to grow. In consequence, there was probably little pressure to run projects profitably. According to various interviewees, FCSP handled money very inefficiently at the time. Project managers had no budgets and virtually no idea of the costs involved with a project. Until 1991, for example, no financial calculations were made per project. Only the R&D department maintained a modest project administration. There was also no interim evaluation of projects according to financial standards.

4.4 QUESTIONS AND HYPOTHESES DERIVED FROM EXPLORATORY STUDY

Chemco's decision in the late 1960s to diversify into fine chemicals was important strategically and intended to make the concern less dependent on economic cycles. But when we examine the implementation of this strategic decision, a few things stand out. In the first place, we can observe that the employment of technological knowledge was never really a problem. On the contrary, Chemco is a highly technology-driven company with a strong technological base. But perhaps the company was too much driven by technology. After all, the market was only discovered by FCSP in the course of the process. The company realised that it was not so effective in a number of marketing aspects. New product proposals were judged in terms of whether the products could be produced and marketed. Although this was not always done as well as it could have been, at least some attention was paid to it.

What this case has to teach us in particular is that the company thought in terms of technological and marketing strengths but hardly considered the organisation to be a tool at all. During the entire process of change – and this is what is so striking – the organisation itself was left virtually unexamined. FCSP tried to be innovative with an organisational structure and organisational routines derived from a commodity business. But these were not at all suitable for achieving innovative success with fine chemicals. The need for a different organisational approach (different from the usual one at Chemco) was hardly ever considered. Some efforts were made in the direction of project management, but they lacked the real commitment from management necessary to make them effective. The early period was characterised by a sort of *laissez-faire* attitude. They were going to undertake something different and had the confidence that they could make a go of it technologically. Organisational aspects were dealt with in a somewhat *ad hoc* manner. They concentrated primarily on the question of whether it would succeed technologically. The implicit reasoning appears to have been that success would then follow almost by itself. Organisational action was taken only when things threatened to go really wrong and the concern was not satisfied with the effectiveness of the innovation processes. But then they turned back to the classical management systems which were not suitable for a specialty chemicals business. Actually, they were choosing an even more explicit commodity structure.

During the whole period we studied, then, no substantial effort was made to develop an organisation which was suitable for the new area of activity. *Ad hoc* solutions followed one after the other. The lesson we draw from this story is that technological effort or positioning won't help much unless a company invests in its organisation. In other words, *companies should not only invest in new products and processes but in renewing their organisation as well.*

Our study made us increasingly aware that companies may not be seeing their organisational capabilities as potential competencies. Perhaps companies think too much in terms of technological and (sometimes) marketing competencies and pay too little attention to building up an organisational competency with respect to innovation. A similar phenomenon can be found in the literature on core competencies. Prahalad and Hamel's (1990) ideas are focused on technological competencies. Yet the innovation literature teaches us that competencies should not be conceptualised through an exclusively technological lens. Technological competencies confer a competitive advantage, but so do marketing competencies (Cooper, 1986; Urban and Hippel, 1989) and organisational competencies (see, for example, Burns and Stalker, 1961). From the FCSP case, it might be concluded that each of these three competencies can be deemed critical for the successful implementation of innovations. The research hypothesis underlying the rest of this study thus reads as follows.

Research Hypothesis
The possibilities which an organisation has to innovate successfully are determined by that organisation's strengths in terms of technological, marketing and organisational competencies.

An operationalisation of the competencies will be presented in the following chapters. For now, the following initial operationalisation will suffice:

- *Technological competencies* Technological competencies revolve around innovative efforts that are deemed distinctive to the firm. They involve the 'platforms' to which the firm is anchored, yet do not preclude it from adopting new technologies and investing in R&D to maintain its technological position. They thus include knowledge and experience of certain specific technologies or processes, the possession of installations to produce these technologies or processes in a commercially efficient manner, and the capacity to adapt rapidly to alternative applications.
- *Marketing competencies* Marketing competencies are centred on customers and enable a firm to incorporate customer needs and preferences into product developments and improvements. They are apparent in co-operative relationships with customers, but also in the extent to which customers are a source of innovation. They imply a body of knowledge about markets and market strategy, a reputation as a good supplier, the presence of networks and co-operation with customers, the availability of sufficient means to keep up with market developments, and so on.
- *Organisational competencies* Organisation is a broad concept and includes not only structure, culture and management style but also the strategy of the company concerned. Organisational competencies represent the structures and culture that promote a common orientation in terms of vision and spillover of know-how as well as the flexibility to experiment and create (diversity of ideas). These attributes characterise the firm's typical project arrangements, interdepartmental relationships and organisational culture. In short, these are the premises of innovative activities that enable the firm to be organic, adaptive and flexible. The point is to have an organisation which promotes innovation.

On the basis of these observations we propose the following hypothesis:

Hypothesis 4.1
A firm's innovative success is determined by the presence of organisational, technological and marketing competencies.

It is important to stress once more that in identifying the technological, marketing and organisational competencies as described above we shall aim to identify the *managerial* technological, marketing and organisational competencies, as was discussed in subsection 2.5.4. We are particularly interested in knowledge about managerial competencies which is transferable to and useful for other companies.

Following the crisis in the late 1970s, the early 1980s were unmistakably characterised by a wave of innovations in the management and organisation of companies. The new management credos penetrated the management and scientific literature, but in most cases were developed in practice, which in many cases happened to be Japanese practice. Consider such elements as total quality management, integral logistics, JIT (just-in-time), integral services, ecological production, concentration on core activities and, most recently, lean production and activity-based costing.

Many firms began industriously to apply these approaches and the consultancy sector made great headway in the process up until the early 1990s. What these management credos have in common, however, is that they are presented as individual therapies, as distinct solutions for organisational problems. In reality, of course, these approaches cannot be separated from each other. Movements of goods are closely connected with the process set-up; product quality is linked with environmental effects; the build-up of technological knowledge cannot be seen in isolation from personnel policy; and customer orientation demands that attention be devoted to the logistical system as well as to training and recruitment. Partial solutions applied to the organisation one by one appear to have only short-term effects (Skinner, 1974). Only an integral approach will provide real help. Translated to the competency model, this implies that innovative success cannot be based on a single competency but requires a combination of core competencies. This observation, however, raises the question as to whether a sort of compensation between competencies is possible. For example, can a firm successfully innovate if it is only strong in two out of three competencies? Questions like these regarding a possible synergy between competencies led to a second research hypothesis:

Hypothesis 4.2
If competencies are jointly present, firms are more likely to innovate successfully.

This proposition has been closely examined in this study. Thus, while the competencies can be broken down into clusters, it is important to recognise their integral role in advancing effective innovativeness. For example, firms with well-developed supplier and customer relationships have better access to external technology (Mansfield, 1989; Urban and Hippel, 1989); firms that make extensive investments in proprietary know-how can better absorb other firms' know-how (Cohen and Levinthal, 1989); and firms with well-established collaborative relationships between R&D and marketing are more likely to incorporate customer needs into the early stages of a product development process (Souder, 1987, 1988).

Obviously, the three clusters of competencies might be related. For example, firms that have crafted enduring, well-established relationships with customers and suppliers are probably likely to have accumulated greater access to external technology as well, which in turn should enhance their proprietary technological abilities. Likewise, well-established inter-departmental relationships will improve the firm's ability to respond to customer demands in a timely fashion. In turn, heavy investment in proprietary development efforts will allow firms to exploit more effectively other firms' innovative outputs, while well-established relationships between marketing, production and R&D will improve the firm's ability to respond to customer needs in a timely fashion and upgrade relationships with their markets. However, the three clusters of competencies are presented as separate, necessary and sufficient antecedents for effective product development.

Based on the case, one may further assume that successfully innovating organisations display continuous development with respect to the three competencies mentioned and that competencies are thus not static quantities. A changing environment may demand different competencies. Thus Chemco had a number of organisational competencies which enabled it to operate successfully on the bulk markets, but the same organisational routines were a hindrance to achieving success on the fine chemicals and specialties markets.

By contrasting firms that differ in terms of innovative success, we might uncover differences in competencies. While competencies as end results of organisational learning are subject to continuous change, they represent the stock of know-how at a given moment in time. This book documents that stock of know-how in order to uncover innovation success factors. Figure 4.1 provides a simple diagram of the tentative theoretical framework derived from this pilot-study research. We propose that innovative success is determined by a combination of technological, marketing and organisational competencies.

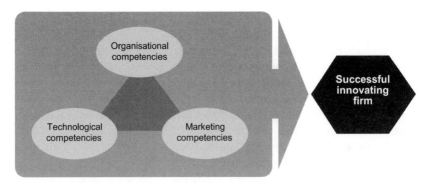

Figure 4.1   The tentative theoretical framework

The case study was an initial step in operationalising the managerial competencies. Three managerial competencies emerged most clearly as building blocks for a framework to explain differences in innovative success. In addition, the case study provided us with two research hypotheses that form the core of our research. Additional hypotheses will be presented in Chapters 5–8 inclusive, but will be derived from findings presented in the literature. The purpose of these hypotheses is to test whether several, mostly single-sector findings from the literature regarding factors explaining differences in innovative success also hold true as non-sector-specific success factors.

PART II

Empirical findings

PART II

Empirical Studies

5. Technological competencies

> R&D means transforming money into good ideas.
> Innovation means transforming good ideas into money.

5.1 THEORY

5.1.1 Technology management

The term *technology* stems from the Greek word *techne*, meaning art. It signifies the knowledge, capacity or skill an individual uses to produce a product or artefact. As used in the modern world, technology refers to the application of knowledge for human purposes (Huxley, 1963). We live in a technology-dominated world where hope is pinned on solving problems through technological advancement. So technology as such will of course have a central place in any study of innovation. Managing technology or, better yet, managing knowledge is at the core of managing innovations. Technology and innovation have always been closely interrelated. With many innovations, the company was primarily guided by its technological possibilities and only to a lesser extent by market needs. And although these so-called 'technology-push' innovations have been losing considerable ground to 'market-pull' innovations, a firm's technological know-how and knowledge still play a crucial role in assuring that the (new) products demanded by the market can be developed and produced. In fact, we can safely state that the innovative potential of a company will be partly determined by its technological knowledge base.

But the mere presence of technological knowledge in a company does not guarantee that it will be used effectively, nor does the possession of technological knowledge as such guarantee sustainable competitive success. Or, as the R&D manager of a front-runner stated: 'Innovation is more than nice research work, it is more than going for a Nobel prize.' Technological knowledge can be regarded as a potential created by, among other activities, R&D, engineering and technology acquisition. Whether or not a company will be capable of exploiting this potential to the fullest is determined by the way it utilises its technological knowledge and manages its technological assets. This involves various strategic questions (for example, the type of technologies to develop, whether to seek technological leadership in these technologies, whether or not to license technology, the size and breakdown of the R&D budget). In past decades, a growing number of companies have recognised that technology is a strategic asset which needs to be managed accordingly. Many companies have come a long way from the leap of faith attitude towards technological developments (Box 5.1)

BOX 5.1 CATEGORIES OF LINKS BETWEEN BUSINESS AND TECHNOLOGICAL STRATEGY DEVELOPMENT

Based on a study of the experiences of a number of companies in various industries, Frohman (1980) identifies the following five categories of links between business and technical strategy development:

• *Leap of faith* Top management calls or allows for certain amounts of money to be spent on R&D without attempting to understand the nature of the specific R&D investment.

• *Lack of faith*   Money allocated to R&D is closely controlled by business management and in some cases business managers are seen to involve themselves in the day-to-day management of R&D projects.

• *Technology-driven* These companies allow their paths of development to be determined by technological breakthroughs. In these companies, which are often dominated by scientists or engineers, business functions respond to the company's latest technical breakthrough.

• *Customer-driven* Companies characterised as customer-driven employ their technical assets for customer application and technical service. The main strategy is to be responsive to market needs. This cash cow strategy generally leads to a short-range focus.

• *Strategic management* This type of linkage entails the development of a business strategy which reflects the organisation's important technological assets and opportunities. The focus is on determining the optimum allocation of technological resources. The strategic balance involves the interaction between the technological assets of the organisation

> and the determination of the business strategy based upon its
> business goals.

The increasing need for a more strategic view of the firm's technological
capabilities and resources is often reflected in the institutionalisation of these
activities as a core business function, a process generally referred to as
'technology management' (or 'innovation management'). It is concerned
with the role of technological innovation in all facets of company
management. This includes, among others:

- selecting and monitoring the core competencies or the technology/
  product/project portfolio (are we doing the right things?);
- planning, developing and implementing technological capabilities in a
  suitable manner (are we doing the right things right?);
- promoting coherence and managing technological knowledge synergy
  among the various business units;
- building and maintaining external relations with other knowledge
  suppliers.

Beside mobilising these internal resources, technology management is also
aimed at external resources. By monitoring long-term trends in relevant
technologies, for instance, new fields of enquiry might be suggested to the
R&D department. Furthermore, technology management is aimed at
orchestrating internal and external resources by creating an appropriate
internal environment that is prepared to learn from the signals the company is
receiving from its external environment. In this respect, it is important to
develop the potential to incorporate knowledge from outside the company, to
spread it internally and apply this newly acquired knowledge in products,
services and processes. In that sense, it is clear that the sphere of influence of
technology management goes beyond that of R&D management (Roussel et
al., 1991; Hertog and Huizenga, 1997). The latter can be regarded as an
element of the technology management function. In fact, since technology
management also deals with make-or-buy decisions regarding technology,
companies with only limited in-house R&D activities might have a very
active technology management function.

The value of technology management is especially recognised by the
large multilevel diversified companies which aim at synergising the
knowledge developed around their core technologies in a variety of divisions
or business units. But it is as relevant to the SMEs as it is to the large
companies, although the scope will differ (Brown and Cobbenhagen, 1998c).
In general, SMEs often have a smaller base of technologies than larger

companies and usually lack possibilities for spreading development risks, making the management of this base even more important. Roughly speaking, we can distinguish two basic types of innovating SMEs: first, the companies which lack internal R&D capacity and have to acquire knowledge outside. They follow external developments and make extensive use of external knowledge. Since they focus on a smaller (sometimes geographical) market, they can do things large companies cannot. This characterises the bulk of the innovating SMEs. In the second category come the very innovative SMEs who develop new products or services themselves. Technology management is as relevant to high-tech firms as it is to low- and medium-tech companies. The scope and emphasis will differ for the various types of companies, but the reasoning behind the technology management function is the same for all.

In our framework, technology management draws heavily on a firm's organisational and technological competencies. Aspects of this business function will therefore be discussed in two chapters. The current chapter emphasises the technologically related aspects of the technology management function as well as organisational issues related to research and technological development. Attention is paid to the scope and advancement of the company's technological knowledge as well as to several strategic choices made regarding its technological developments. The general managerial and organisational aspects of innovation management will then be discussed in Chapter 7 in connection with the analysis of the organisational competencies.

The technological and organisational competencies as utilised in this study are related to this broad technology management function. In this book, we examine the form, not the content, of the competencies. These competencies are constructed on the basis of the routines by which knowledge is developed, channelled and used. These routines determine what the company is able to do (potential behaviour). Yet a company can only develop competencies when the routines are mastered and potential behaviour is transformed into actual behaviour. So technological competencies can be regarded as a firm's proficiency in applying knowledge in such a way that a competitive advantage is gained.

In this study, we are thus particularly concerned with the manner in which the organisation 'deals' with technological knowledge and skills. We do not intend to evaluate the technological knowledge itself. Since our aim is to study non-sector-specific success factors, we are only interested in general competencies and not in company- or sector-specific technical competencies or specific technical knowledge. As a comparison of technologies mastered by companies is thus outside the scope of this study, it is possible to draw causal relationships between the kinds of technologies a company masters and its innovative success. What can be studied are characteristics such as the breadth or range of the technologies mastered by a company as well as a general self-appreciation of the sophistication of the company's technological state of the art. Consequently, the discussion concerning the companies' technological capabilities will be conducted at a higher conceptual level. The focus in this book is thus on the way technological knowledge is handled in the company, as this is a managerial competency.

5.1.2 Research and development expenditures

Companies usually invest in research and development (R&D) with the aim of developing or maintaining a competitive advantage. Effective R&D aimed at product innovations can, for instance, lead to successful new products, which in turn might lead to an increase in turnover, higher market shares or even increased profit. R&D aimed at process innovations can result in lower production costs, increased product quality, and so on. The scientific literature, however, is not conclusive in determining links between R&D investments and company success. Research among companies in a variety of industries has shown associations between R&D spending and subsequent growth in sales, but no clear signs of R&D expenditure ties to profitability have been found. Morbey (1988), for instance, found that companies which invested a larger percentage of sales in R&D in 1976 benefited from a greater growth rate in their industry (between 1976 and 1985) than their competitors. For the sample studied, a minimum R&D level of 3 per cent of sales revenue was identified as the critical level for ensuring long-term growth. However, Morbey could not detect a strong correlation between R&D intensity and growth in profitability. A similar conclusion was drawn from a study of 50 prominent hardware, software and networking companies (Anthes and Betts, 1994). This study also failed to show correlations between investments in R&D and short-term profits, nor could company success be attributed to R&D expenditures as such. There are just too many other intervening variables which make it virtually impossible to prove direct linkages between technology measures, such as R&D intensity, and company profitability (Roberts, 1995). So, although it can be argued that R&D expenditures and innovative success are related, the one does not necessarily have to result from the other. R&D expenditures are just one of many variables which might account for sales growth and rising profits. Heavy spending on R&D thus seems to be no guarantee of overall corporate success.

None the less, the level of R&D expenditures is frequently employed as an indicator of a company's technological activities. This yardstick has a significant drawback, however, in that it says nothing about the productivity of R&D efforts and takes virtually no account of the company's other technological activities (such as production engineering, design and software development). Thus, the absolute level of the R&D ratio does not provide sufficient insight into the extent to which technologies are applied. It tells us nothing, for example, about the state of the art of the technologies employed (which may have been acquired by purchase or may be incorporated in machines). Basically, R&D figures do not provide any answers to the two crucial questions: 'Are we doing the *right things*?' and 'Are we doing the right things *right*?' In fact, the manner in which the R&D budget is allocated is probably more significant than the R&D expenditures themselves, since it involves, among other things, strategic choices of research areas. Some support for this hypothesis is given by Roberts (1995), who found that company sales growth was statistically related to overall R&D managerial capability in a sample of 244 of the largest companies conducting R&D in the US, Western Europe and Japan.

It should be emphasised that our study is aimed at identifying factors which account for innovative *success*, not just for innovativeness (as defined in Box 3.3). An innovative company is a company which has a track record of introducing new products on the market or incorporating new technologies in its processes. A successfully innovative company is an innovative company which can also exploit its innovativeness successfully in commercial terms. While the relation between investments in R&D and *innovativeness* might be obvious, it is not the relation we are studying here. Instead, we focus on the question whether factors like the size of the R&D budget are related to the success of the innovative efforts. Based on the literature, we are prone to state that the size of the R&D budget in itself is not an all-determining condition for innovative success. We therefore hypothesise as follows.

Hypothesis 5.1
   Successfully innovating companies do not invest significantly more in R&D than less successfully innovating companies.

5.1.3   Internal technological knowledge

The sophistication of the technologies a company masters can range from 'state-of-the-shelf' public knowledge to proprietary knowledge that no one else has (Leonard-Barton, 1991b). The former type of knowledge allows a company to maintain its position, avoid deterioration of current processes or even catch up. In some cases, it can even provide a company with a competitive advantage. For example, if the company is capable of combining public knowledge in unique and unprecedented ways (the Sony Walkman, for instance, involved a combination of miniaturisation technology and stereophonic tape technology, both of which were readily available at the time). However, chances are great that these companies will soon be imitated. Proprietary technology, on the other hand, makes it possible to leapfrog the competition if the investment is significant enough to impact a whole line of business (ibid.). This, however, is an extreme and infrequent benefit. But between these two extremes a continuum of various levels of technological sophistication can be discerned. A strategic question for every innovating company concerns the degree to which the firm desires to be a technological leader. Companies wishing to maintain an advanced level of technology may decide against a leadership role in favour of a 'me too' strategy (very rapid imitation) and/or a 'second but better' strategy (imitation with clear improvements). One can point to successful cases for each of these strategies (or modified versions). For years, many Japanese companies were adepts of the latter strategy. At the time, Sony formed the exception by opting for technological leadership, but it ceased being exceptional in this respect some time ago. Nowadays, many Japanese companies are technological leaders.

A relation between technological advancement and company success has been documented by Meyer and Roberts (1988). They have shown that companies which lacked technological aggressiveness and undertook only minor improvements to their core technologies performed less well than companies which had made major enhancements in core technology over the years. We therefore propose that front-runners are technologically more advanced (and thus more likely to innovate at the forefront of technological developments) than pack members.

The increasing attention being focused on the development of systems and the increasing complexity of products call for technological multidisciplinarity (the ability to combine technologies). Often this implies that companies must master a broader set of technologies. In this respect, we presume that front-runners outperform pack members in terms of the range of technologies mastered. The former will be more broadly orientated than the latter. Summarising, we hypothesize as follows.

Hypothesis 5.2a
> **Successfully innovating companies (front-runners) are more sophisticated technologically than pack members.**

Hypothesis 5.2b
> **Front-runners have a broader technological scope than pack members.**

5.1.4 Capabilities for incorporating external knowledge

Hypotheses 5.1 and 5.2 deal with the company's internal technological knowledge. But apart from the knowledge stored and developed inside their own organisations, most companies have to tap into external sources of knowledge as well. Innovation is not a narcissistic exercise which companies must perform entirely solo. Instead, they depend on their environments for knowledge and ideas. R&D, particularly the exploratory aspects, is more and more frequently contracted out. In many sectors, the required disciplines for long-term research are becoming more and more diverse, and it is not possible to have all the necessary knowledge (and equipment) in-house. Knowledge transfer with other companies and institutes is therefore becoming increasingly important. Thus, as was stated in subsection 5.1.1, technological competence is also reflected in the ability to deal effectively with know-how from other organisations. And this brings up the question of how to tap into this external knowledge and translate it into internally usable knowledge. The external dimension of technology management is therefore concerned with the creation of an appropriate internal environment that is prepared to learn from the signals received by the company from its external environment. It is crucial in this respect to make optimum use of internal and external sources and channels of information during the innovation process. This requires profound insight into the technological developments taking place outside a company's own environment.

The external environment is not only a source of technological knowledge, but also provides companies with a variety of sources of ideas for the development of new products and processes. Contacts with suppliers have been shown to account for the superior innovativeness of Japanese automobile manufacturers (Clark and Fujimoto, 1991), while end users often provide significant impetus for product or service improvements (see, for example, Urban and Hippel, 1989).

The pressure on many companies to innovate in combination with ever-shortening product life-cycles puts increasing pressure on their R&D departments to operate more efficiently. It is often cheaper to use knowledge developed elsewhere than to develop this knowledge internally (Box 5.2). More distinct choices are being made, evaluations are becoming more frequent and stringent and firms are more active than they used to be in prospecting for knowledge outside the company. Technology transfer is thus becoming an important instrument for companies that want to keep their R&D affordable. Bearing in mind that our study is focused on the medium- and low-tech, medium-sized company, it is conceivable to assert that these companies will generally not be equipped with abundant R&D facilities and will therefore have to rely even more on the acquisition and transfer of technology generated elsewhere.

BOX 5.2 FIRST SEARCH, THEN RESEARCH!

In a keynote speech at a conference on knowledge transfer, Dr Ir Beckers (former vice-president of Shell R&D) illustrated how far management can go with this strategy. He reported on his recent visit to a Japanese company. The company's R&D lab was a large open-plan office where the researchers, their department heads and the director all had their desks. They were in continuous contact with each other. The floor was surrounded by a glass wall, behind which a few isolated individuals in white coats did their work. The philosophy underlying this arrangement was that researchers should first search for knowledge available in the literature and patents before beginning research themselves. If the required knowledge could not be obtained from these sources, they would have to contact universities and research centres all over the world. Every conceivable source of knowledge would have to be tapped. Only when all other means had failed would the most costly alternative be resorted to: doing one's own R&D. Thus the motto: First search, then research!

The question, however, is whether a company's external orientation is a distinguishing factor explaining innovative success. Do successfully innovating companies rely mainly on internally generated knowledge (because they innovate at the forefront of technological knowledge, for example) or is externally generated knowledge a crucial input as well? We expect the latter. Thus, when looking for non-sector-specific innovation success factors we hypothesis that successfully innovating companies will make significantly more use of external sources than less successfully innovating companies. Nevertheless, empirical studies have not been very coherent in linking innovative success to the intensity of use of external and internal sources of knowledge. A review of eight different studies on the use of sources of information during the innovation process (Meyer, 1985) showed that the studies differed to a large extent in pinpointing the proportion of externally generated information messages to the total number. These ranged from 11 per cent to 65 per cent of the total number of information messages. Some studies (Ettlie, 1976) therefore suggest that the various information channels are equally important but are used differently at different times, whereas others (Olsen, 1975; Von Hippel, 1982) assert that the main sources of innovation might be dependent on the company's industrial sector. Although it is extremely difficult to compare these studies (different sample sizes, different sectors, different selection methods, and so on), we can at least conclude (with our non-sector-specific perspective in mind) that there is great ambiguity in the literature concerning the use of external sources.

The networking organisation improves its competitiveness by externalising some of its innovative efforts and developing interfirm relationships. Apart from co-operation focused on knowledge transfer, companies can also engage in relationships in order to share the risks and costs of (often complex) innovations. As we hypothesise that front-runners are more broadly orientated and advanced technologically than pack members, we may assume that this calls for greater co-operation with external partners as well.

In defining the external orientation of a company as the extent to which it searches for knowledge outside and involves external partners in the innovation process, we hypothesise as follows.

Hypothesis 5.3a
   Successfully innovating companies are more externally orientated than pack members.

Levinthal (1992) argues that a company's research activities must both generate products and enhance its ability to learn. This ability to learn is called the firm's 'absorptive capacity' and defined as 'the ability of a company to recognise the value of new external information, to assimilate it and apply it to commercial ends' (Cohen and Levinthal, 1990). A firm's absorptive capacity depends on its interfaces with the external environment as well as on the transfer of knowledge between and within its subunits. According to Levinthal (1992), a firm's absorptive capacity may be its most fundamental core technological competency. To enhance their absorptive capacities, corporations must learn to transfer knowledge between organisational functions and between organisations (for instance, through co-operative research efforts). In this manner, the firm's capabilities become leveraged. The organisation's ability to tap into externally available technological knowledge in a prompt and active manner is thus crucial for building up its absorptive capacity. We therefore presume that successfully innovating companies will be prompt in detecting new technological developments and will often be among the first to adopt a new technology to incorporate in their products and/or processes. In other words:

Hypothesis 5.3b
   Successfully innovating companies are more proactive in tapping into externally developed knowledge than are pack members.

5.2 RESULTS

5.2.1 R&D management

In order to test whether successfully innovating companies invest significantly more in R&D than less successfully innovating companies (Hypothesis 5.1), we asked the companies to provide detailed R&D figures. Not all companies were able or willing to provide us with this information. As average R&D ratios differ widely per industry, the matched-pairs principle must be applied very consistently with respect to a comparison of such figures. For that reason, we have decided to use only the 22 pairs of which both companies provided us with R&D data for analysing R&D ratios. An examination of these 22 full pairs reveals that the average R&D ratios (R&D expenditures as a percentage of sales) of front-runners do not differ significantly from those of pack members. The average front-runner invests 2.3 per cent of its annual turnover in R&D, while the average pack member invests 1.6 per cent. At face value this difference amounts to some 40 per cent. That seems like a lot but, partially due to the very great spread, it does not turn out to be statistically significant ($t = -1.30$, $p < 0.2$). What is more important is the observation that the companies do not seem to attach great value to these hard R&D figures. In fact, it was quite remarkable that many companies did not have this information available or could only produce it with great difficulty. Many companies could not provide us with these figures during the interview, but mailed them afterwards.

Based on the pilot questionnaire, however, we had foreseen this problem. Therefore a comparison of R&D expenditures was provided for in a different manner in the questionnaire: as a relative (more qualitative) comparison of the R&D percentage with that of the sector. Thus, each company was also asked to compare its R&D spending to the industry average on a five-point scale ranging from 'well below' to 'well above'. It then emerges that the front-runners can generally be regarded as more R&D intensive than the industry average. Only 17 per cent of all companies were below industry average in this respect, while 21 per cent were around the industry average and 62 per cent above. Of the latter, almost half even scored 'well above' industry average. When comparing the front-runners with the pack members, we see a fairly strong correlation between the relative size of the R&D ratio and innovative success. On a five-point scale, pack members scored on average 3.1 (industry average) and front-runners 4.1 (above industry average). The T-test proved to be highly significant ($t = -3.73$, $p < 0.0001$).

Furthermore, the R&D ratio of most front-runners increased throughout the five years preceding the interview, while that of most pack members remained the same or in some cases declined slightly. We may thus conclude that there are strong indications that front-runners spend more on R&D than pack members. Hypothesis 5.1, which postulated that successfully innovating companies do not invest significantly more in R&D than less successfully innovating companies, might therefore be rejected.

We have also studied the breakdown of the R&D budget to learn more about possible differences in R&D spending between the two groups; for instance, differences in expenditures on short- and long-term research or differences in the allocation of R&D spending on product versus process innovations. Cross-tab analyses revealed no such indications (see also Tables 5.1a and 5.1b). The distribution of the total R&D budget over the diverse types of activities differs hardly at all between front-runners and pack members.

Table 5.1a Distribution of R&D expenditures over type of R&D (%)

	Pack members	Front-runners
Developing new products	29	33
Improving existing products	27	28
Discovering new uses for existing products	6	7
Lowering production costs	17	13
Service to sales or production	13	10
Basic research	3	5

n = 46 companies (23 complete pairs)

Table 5.1b Distribution of R&D expenditures over type of innovation (%)

	Pack members	Front-runners
Product innovations	68	61
Process innovations	29	33
Unknown	3	6

n = 46 companies (23 complete pairs)

5.2.2 Technological advancement

Both the R&D manager and the general manager indicate that front-runners are technologically only slightly more sophisticated than pack members.

Technological sophistication or even leadership in a sector is not exclusively reserved for front-runners. Slightly more than half of the pack members consider themselves to have a leading technological position in the sector as against three-quarters of the front-runners. About two-fifths of the pack members regard their position as 'average' and only two pack members state that they are lagging behind their competitors (Figure 5.1).

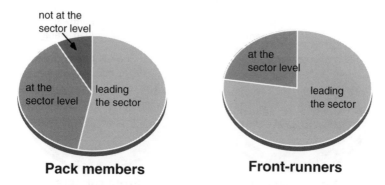

Pack members **Front-runners**

Figure 5.1 Technological sophistication

The T-test shows that these differences are not significant. These data provide some support for the hypothesis that front-runners are technologically more sophisticated than the average company in their sectors, but they also indicate that the same is true for pack members. Technological sophistication can hence hardly be regarded as a distinguishing characteristic between front-runners and pack members. Hypothesis 5.2a can therefore not be supported.

However, we did find a close correlation between technological sophistication and the firm's innovativeness ($r = 0.46$, $p < 0.01$). Companies with a strong technological position can be further characterised as companies which are prompt in picking up new technologies ($r = 0.40$, $p < 0.01$); creative ($r = 0.41$, $p < 0.01$) and technology-driven ($r = 0.30$, $p < 0.05$). Moreover, they appear to have a higher R&D ratio ($r = 0.33$, $p < 0.01$). Apparently, innovativeness and technological advancement are more closely related than innovative success and technological advancement.

According to Hypothesis 5.2b, front-runners should have a broader technological base than pack members. There do appear to be some indications for this here, but they are not statistically significant. 72 per cent of the front-runners, state that they have mastered a wider range of technologies than their competitors and only 3 per cent find that they have mastered a narrower range of technologies than their competitors (Figure

5.2). The proportion of companies among the pack members which say that they have mastered a broader than average range of technologies is smaller (only 57 per cent), and 14 per cent of the pack members state that the range of technologies they have mastered is narrower than their competitors. The T-test shows that these differences are not significant ($t = -1.51$, $p = 0.07$). These data thus provide no conclusive support for Hypothesis 5.2b.

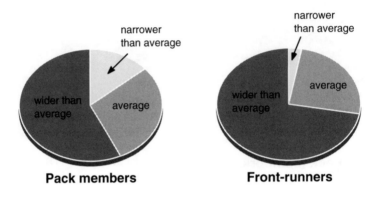

Figure 5.2 Breadth of range of technologies mastered

In order to come up with a more stable measure of technological leadership, a scale was constructed out of the variables indicating the firm's relative technological capabilities, its technological sophistication, the breadth of the range of technologies it has mastered and the extent to which it can be regarded as a trend-setter (Table 5.2). Cronbach's alpha of this scale (called 'technological leadership') is acceptable (0.7). Front-runners appear to score better on this scale than pack members. The T-test shows a significant difference ($t = -2.42$, $p < 0.01$). Thus, although the front-runners do not always differ significantly from the pack members on individual variables measuring aspects of technological leadership, they do differ significantly on a scale derived from these variables. Hypotheses 5.2a and 5.2b, which presume close relations between, on the one hand, technological sophistication and technological scope and, on the other hand, innovative success, can therefore neither be rejected nor accepted. We may thus assume that it is the combination of capabilities, technological breadth and sophistication as well as the fact of being a trend-setter which allows front-runners to differentiate themselves from pack members.

Table 5.2 Scale 'technological leader' constructed from individual variables

Scale	Scales or variables used to construct the scale	Cronbach alpha	Correlation with innovative success
Technological leader		0.7	0.46***
	Technological capabilities		
	Technological breadth		
	Technological sophistication		
	Trend-setter		

Notes: *** $\alpha \leq 0.001$ (one-tailed)

5.2.3 External orientation

A firm's external orientation regarding R&D is measured by the extent to which R&D is outsourced, the value which the company attributes to external sources of innovation and the extent to which it co-operates with outside parties during innovation processes.

Outsourced R&D
An analysis based on data from a subsample of complete pairs should shed some light on the extent to which the two groups differ with respect to outsourcing R&D. Only 14 complete pairs (28 companies) provided us with data about the man-years spent on R&D activities (Table 5.3). The impression we received from this analysis was that front-runners spent more full-time equivalents (fte) within their R&D departments and more on R&D-related activities in other company departments than pack members. These differences were statistically significant. Although at first glance the difference regarding outsourcing is quite large, it is not significant. The main reason for this is that the distribution is very uneven and the sample size is fairly small (28 companies). The high score for the front-runners is heavily dominated by one front-runner which outsources 46 fte on R&D activities. Furthermore, only seven of the 28 companies in this subsample say that they have spent R&D money outside the company. This number of observations is too small to base generalised conclusions on.

Sources
The respondents were asked to indicate what value they attributed to various internal and external parties as sources of product innovation, process innovation and technological development for the company. These parties

were then categorised as either internal or external and their average importance was calculated. Since these figures are averages of lists of internal and/or external sources, it is not possible to compare the average value assigned to the internal sources with that assigned to the external sources. As Table 5.4 shows, there are no significant differences between front-runners and pack members regarding the orientation of sources. There are, however, some slight, yet not significant, indications that front-runners are more externally orientated than pack members with respect to technological developments.

Table 5.3 R&D expenditure in terms of full time equivalents (fte)

Fte on R&D related activities	Pack members	Front-runners	T-value
Spent in R&D department	1.6	3.9	–1.81*
Department other than R&D department	0.6	2.2	–2.13*
Outsourced (outside the company)	0.4	3.4	–0.92

n = 28 companies (14 complete pairs)

Notes: * $\alpha \leq 0.05$; ** $\alpha \leq 0.01$; *** $\alpha \leq 0.001$ (one-tailed)

Table 5.4 Average importance of internal and external sources: pack members versus front-runners *

	Pack members	Front-runners	T-value
Product innovations			
Internal sources	2.66	2.60	0.60
External source	2.43	2.45	–0.19
Process innovations			
Internal sources	2.57	2.72	–1.24
External source	2.26	2.25	0.03
Technological development			
Internal sources	3.05	3.19	–0.79
External source	2.11	2.33	–1.57

Note: * 1= unimportant; 2= somewhat important; 3= important; 4= very important; 5= crucial

To enable us to compare the relative importance of the various internal and external sources, the internal and external categories have been broken down into the individual sources in Table 5.5. This table lists the five most important sources and the value attributed to them for each type of innovation.

Table 5.5 Main sources for three types of innovation

	Pack members			Front-runners		
	Source	Value	Type	Source	Value	Type
Product innovations	Marketing	3.8	Internal	Customers	4.0	External
	Customers	3.7	External	Marketing	3.7	Internal
	R&D	3.4	Internal	R&D	3.3	Internal
	Management	3.4	Internal	Management	3.3	Internal
	Competitors	3.2	External	Competitors	3.1	External
Process innovations	Management	3.3	Internal	R&D	3.2	Internal
	Production	3.2	Internal	Engineering	3.2	Internal
	Literature	2.9	External	Management	3.1	Internal
	Engineering	2.8	Internal	Production	3.1	Internal
	R&D	2.8	Internal	Marketing	2.8	Internal
Technological developments	R&D	3.8	Internal	R&D	4.1	Internal
	Publications	3.3	External	Publications	3.2	External
	New employees	3.1	Internal	New employees	3.0	Internal
	Customers	3.0	Internal	Customers	2.8	Internal
	Suppliers	2.7	External	Technology fairs	2.7	External

With regard to the sources of *product innovations*, front-runners hardly differ from pack members. Both groups have a strong market-pull orientation: the customers and the marketing department are the most important sources of product innovation. Internal R&D is rated third in importance. Next come the management team and competitors. Despite the fact that marketing is formally an internal source, it is very much focused on information from the market, so we might state that the two most important sources for product innovation are externally orientated.

For *process innovation*, the most important sources are internal. Furthermore, the differences between front-runners and pack members are much larger than is the cases for product innovation sources. It is remarkable, for example, that pack members indicate their own management teams and production departments as the most important sources of process innovations. For front-runners, these two sources are in third and fourth place respectively. R&D and engineering are the most important sources for them. The correlation of a factor-derived scale averaging the scores on R&D and engineering as sources of innovation with innovative success is low (0.27), but significant at the $p < 0.05$ level. From this, it may be cautiously concluded that front-runners anticipate process developments more than pack members do. After all, the most important ideas come from two departments which are not directly involved in the daily production activities. With pack members, process innovations seem to be initiated through experiences with the current process (with the production department and the management team as the most important sources) rather than through developments outside the company.

With respect to *technological developments*, front-runners do not differ much from pack members in the value they attribute to the various sources. The companies in the sample appear to develop primarily on the basis of their own knowledge and skills. For both groups, the most important source of technological development is internal R&D (in both the R&D department and other company departments). Next come publications, hiring new employees with the required know-how, and knowledge transfer from suppliers, customers and technology fairs. However, the difference in value attributed to internal R&D (most important source) and publications (second most important source) respectively is about twice as much for front-runners as for pack members. Other ways of bringing technologies which are 'foreign to the company' into the organisation, such as joint ventures, takeovers, purchase of licences and patents and co-operation with competitors, appear to score very low as sources of technological development.

External co-operation

The third measure of external orientation is the extent to which the company is involved in co-operative relationships with outside parties. It was tested whether the two groups differed with respect to their technological co-operation with suppliers, competitors, intermediary organisations, research institutes and universities.[1] Both groups say that they co-operate quite often with suppliers (Table 5.6). In fact, suppliers are generally the most important external partner for both groups. With respect to the other partners, there are some indications that pack members tend to co-operate more with competitors than front-runners, while front-runners tend to co-operate more with non-competitive research facilities such as research centres and universities. However, most of these outside partners are only seldom involved.

Table 5.6 Co-operation with partners

	Pack members	Front-runners	T-value
Suppliers	3.35	3.50	−0.74
(Semi-)public research centres	2.21	2.46	−1.08
Universities	2.02	2.40	−1.55
Pre-competitive co-operation with competitors	2.22	2.11	0.53
Regional technology advisory centres	1.76	2.01	−1.27
Competitors after market introduction	2.05	1.86	0.95

Notes: Partners are ranked in order of importance that the average front-runner attributes to this form of co-operation. * $\alpha \le .05$; ** $\alpha \le .01$; *** $\alpha \le .001$ (one-tailed)

When we examine the co-operative relationships with suppliers in greater detail (Table 5.7), it becomes clear that the most important reason why both groups involve suppliers is as sources of ideas for process innovations and partners in developing these innovations. Suppliers are only marginally involved in the development of new products and services.

Table 5.7 Type of co-operation with suppliers

	Pack members	Front-runners	T-value
Process innovation ideas	3.36	3.26	0.31
Process innovation development	3.36	3.00	1.13
New product ideas	2.71	2.85	−0.46
New product development	3.11	3.00	0.33

Notes: Partners are ranked in order of importance the average front-runner attributes to this form of co-operation. * $\alpha \le .05$; ** $\alpha \le .01$; *** $\alpha \le .001$ (one-tailed)

Based on the quantitative data, we cannot state conclusively that front-runners are more externally orientated than pack members with respect to product and process innovations (as Hypothesis 5.3a proposes). According to some measurements they are, but according to others they are not. To test the proposition regarding the technological absorptive capacity of companies (Hypothesis 5.3b), a scale was constructed from various variables which measured certain aspects of the company's technological absorptive capacity

(Table 5.8). This scale (Cronbach's alpha $= 0.7)^2$ proves to be closely correlated with innovative success (T $= -2.73$, p < 0.005). The variables used to construct the scale are the variables measuring the extent to which the company detects technological and market developments promptly; the extent to which the company is regarded as being prompt to adopt new technologies; the ease with which new technologies are used to reduce production costs (relative to the sector average); and the extent to which the company benefits from technology transfer (relative to its competitors). The technological absorptive capacity scale thus provides strong support for Hypothesis 5.3b.

Table 5.8 Scale 'technological absorptive capacity' constructed from individual variables

Scale	Scales or variables used to construct the scale	Cronbach's alpha	Correlation with innovative success
Technological absorptive capacity		0.7	0.51 ***
	Sensing new technological developments promptly		
	Prompt to adopt new technologies		
	Active in technological knowledge transfer		
	Leading in terms of cost reductions through new technologies		

Notes: * $\alpha \leq 0.05$; ** $\alpha \leq 0.01$; *** $\alpha \leq 0.001$ (one-tailed)

5.2.4   Technological competencies and innovation success

To find support for Hypothesis 4.1, which stated that technological competencies are important predictors for innovative success, the significantly correlating variables were used to construct a multiple regression equation. A stepwise multiple regression analysis was employed, which allowed us to limit the number of variables in the final equation to a minimum. To facilitate a comparison of the regression coefficients of the variables, it was decided to compute the regression model with the normalised values. Table 5.9 shows the effects of technological competencies on innovative success. The technological competencies, as operationalised by this model, appear to explain 41 per cent of the variability in innovative success.

Table 5.9 Multiple regression analysis of effect of technological competencies on innovative success

Independent variable	B	Beta	t
Relative R&D expenditures	0.088	0.298	2.47**
Innovativeness	0.076	0.242	1.80*
Technological absorptive capacity	0.062	0.230	1.80*
Constant	0.624	0.026	24.02***

$R^2 = 0.41$, $F(3,59) = 13.40***$

Notes: * $\alpha \leq 0.05$; ** $\alpha \leq 0.01$; *** $\alpha \leq 0.001$

While all of these variables correlate significantly with innovative success, it is primarily the variable that indicates the relative R&D expenditures that is significantly related to innovation success in the regression analysis. The scale measuring the company's technological leadership failed to reach the level of significance required for inclusion in the regression equation. This is largely due to the high multicolinearity among the variables. The 'technological competencies' can thus be operationalised through the innovativeness of the company; the relative size of its R&D budget as opposed to that of its competitors; and the extent to which the company is capable of transforming externally available knowledge and information into internally useful knowledge.

The results demonstrated a coefficient of determination (R^2) of 0.41, suggesting that while technological competencies are an important predictor of innovative success, they are by no means exclusive.

5.3   DISCUSSION

5.3.1   R&D figures: how useful are they?

Technology is an important source of innovative success. Investments in technology development nurture the renewal of products, processes and services. It is a way to obtain a competitive advantage. But technological competencies do not result just from the energy companies invest in research and development. Although the study indicates that front-runners invest more in R&D than do pack members, the differences are only slightly significant. A remarkably large number of companies had no clear idea of their R&D figures and had to resort to estimates. The least reliable are most certainly the R&D ratios from the pack members, as only 30 per cent of them claimed to have a separate R&D budget. Front-runners tended to have more detailed insight into their R&D figures than pack members: 62 per cent of this group said they had R&D budgets; and this difference is highly significant. This is an interesting observation as it indicates that front-runners appear to have much better insight into their R&D spending than pack members.

In scientific and policy circles, great value is generally attached to R&D figures as indicators of company innovativeness. What emerges rather strikingly from our interviews, however, is that the small and medium-sized firms are not all that involved with R&D. Often, they don't even know the extent of their R&D expenditures, or can only produce these figures after long searching. Even less do small and medium-sized companies ask themselves how large the R&D percentage should be. For 'What does an R&D ratio actually tell you?', as many respondents pointed out. After all, an R&D percentage only shows investments in R&D. It says nothing about the efficiency with which the money was spent or the effectiveness of the output. It appears that discussions about R&D are conducted differently in small and medium-sized firms (see Box 5.3). Such companies are often concerned with the technological risks involved. In service companies and companies which work on a project basis, for example, we see that R&D investments are closely related to the risks faced by the company in connection with customer orders. The fact that front-runners are much more aware of their R&D figures than pack members is an indication that knowing what the company is doing in terms of R&D is definitely important. Much more important than the R&D figures themselves is a firm's ability to pick up new developments at the right time. One could call this the antenna function. It has to do primarily with picking up weak signals, knowing that one is more or less involved with R&D than one's competitors, knowing where knowledge can be acquired, knowing what the competition is up to, and so on. The methodology whereby we ask companies to compare their own innovation expenditures with those of the sector is thus quite appropriate. This formulation appears to be more in keeping with the companies' own environments than the question about R&D percentage.

The 'fill in the blanks' exercise with the R&D figures also taught us that not much value can be attached to the R&D figures from small and medium-sized low- and medium-tech firms. Research based on these figures will be built on quicksand. The many cases in which the researcher tried to establish this figure in collaboration with the director or R&D manager are illustrative in this connection (Box 5.4). In fact, R&D percentages are only useful measuring instruments for a restricted range of firms (large companies). Furthermore, the innovation costs of many companies amount to much more than the R&D expenditures alone. They may also include the costs of new or renovated facilities and equipment, marketing, advertising, and so on.

As was said before, the size of the R&D ratio in itself is not a competency, nor is investment in R&D. Technological competencies are more than a simple price tag put on R&D. Rather, the competency is acquired by the way the money is spent, which is a matter of behavior with respect to the development and use of knowledge. Thus, when we look at the story behind the R&D figures, the differences between front-runners and pack members become much greater. The qualitative data give the impression that front-runners can be said to spend their R&D money more 'wisely' than pack members. The development of new technologies, products and processes is by nature a costly and risky endeavour. Not all companies have the financial means or manpower to embark on such an adventure. They have to be smart: smart in searching for useful, already developed technologies in order to translate this knowledge into internally useful information; or smart in investing only in technologies which allow them to produce products with which they can distinguish themselves from their competitors and which are also closely related to market needs. This is especially the case for the small and medium-sized, low- and medium-tech companies studied here.

BOX 5.3 A DIFFERENT WORLD

Front-runner 21 is very profitable. The director is very clear about the investment policy: 'no major investments in bricks and steel'. The most important competitor (the pack member) has recently invested in a state-of-the-art high-tech plant, but nearly went bankrupt a few years ago. The pack member is a typical example of a company which innovates considerably but often does so for the sector as a whole. That is, it appears to be incapable of keeping its innovations exclusive, so that many of its developments are soon taken over by competitors. The company is definitely innovative, but lacks the shrewdness to exploit these innovations commercially. Ironically enough, its competitors are successful in doing so.

A high-tech engineering company (front-runner 39, employing 75 highly skilled engineers) makes advanced systems and is a world leader in some of its market niches. Despite the many new products and processes which it has developed, it does not consider all of them to be innovations. A project is considered to be an innovation only when the company has to invest in knowledge development and run risks with technologies which are new to it. Innovations are the risky projects: projects demanding investments which are difficult to assess in advance in terms of the possibility of finding solutions. Such investments have budgets which can serve as safety nets for the project managers. These are their R&D budgets.

Company 34 is the pack member in this sector. Before entering new markets, this company first attempts to acquire name recognition by means of a so-called pilot project. Such an initial project is seen as an investment which yields no profit but only generates costs. Based on the knowledge, experience and contacts built up in this project, other customers are subsequently sought. Much attention was devoted to technological development in the past since this forms the basis of the company's other activities. With the technological knowledge available within its own organisation, the company tries to keep ahead of the competition, particularly in the field of installation. Moreover, the company has also begun to function in a more market-orientated manner, with the result that it initiates fewer developments itself. It has realised that the idea that it has to develop everything itself was too predominant. Or, as the new director puts it, 'This isn't a playground any longer.' In the past, more development was allowed to take place than the company could actually permit itself. This situation must continue to change.

An accounting firm (front-runner, case 30) sees its R&D budget as a safety net for risky projects which require knowledge development.

BOX 5.4 TOGETHER WE WILL FIND A WAY

The researcher asks the director how high the R&D percentage is. 'That is a good question,' the director responds, 'but I can't answer it just like that. Look, R&D and innovation are involved in so many things. We don't have a separate balance sheet for it or anything like that. We have a development department which is concerned with such matters, but it would be going a bit far to call everything they do R&D.'

Let's try a heuristic approach, thinks the researcher, and offers a helping hand. 'Do you perhaps have an idea of the average R&D ratio in your sector?' He does indeed, for it was recently published in a report. With a little searching, he found the article. The average R&D ratio was 2.1 per cent. 'Okay, now if we look at your company, is it above or below average?' That was an easy one. The director was sure that the firm was below average, but how much he couldn't say. The researcher explains which activities we consider to be R&D expenditures and which we do not. But it is still difficult for the director to come up with a concrete figure.

Subsequently, the researcher goes through the categories of R&D expenditures with the director one by one and asks him to make a general estimate of his company's expenditures. When these were finally added up and divided by the sales figure, the result was an R&D ratio of 1.7 per cent, which, while indeed lower than average for the sector, was none the less not as low as the director actually thought. When the figures were finally in front of him, he concluded, 'Gee, how interesting. Never thought about it that way. This starts me thinking, for we are not doing that well in terms of innovation.'

The front-runners include companies that are able to pick up new technological developments from other areas of application and translate them into their own products, services and processes. Their know-how is often obtained from afield (Box 5.5). This group includes, for example, companies which were heavily involved in collective (precompetitive) research and development activities at the sector level. They were usually able to help guide such developments and profit from them because they had positioned themselves close to the source; for example, by trying to make their norms (with respect to the environment, safety, health, and so on) more stringent on the basis of their own competencies so that other firms would find it more difficult to imitate them. While such results were of course available to all companies in the sector, companies which were 'close to the source' could take advantage of them more rapidly.

BOX 5.5 CRUISE MISSILE TECHNOLOGY IN A LOW-TECH INDUSTRY

Company 25 produces foods in glass which must, of course, be free from pieces of glass and other irregularities. Until recently, there was no detection equipment that was sufficiently reliable within the scope of the norms set by the company. Thus, the final check took place manually. On the initiative of its director, the company started to develop a piece of equipment which would be able to do this job. While looking for technologies which would enable accurate detection, they crossed the path of a supplier of production installations. Eventually, through the American defence industry, they tracked down an image-processing technology which is also applied in cruise missiles. An Italian company subsequently produced an initial prototype on the basis of the available know-how. Company 25 is the first in Europe to use this machine.

5.3.2 External orientation

Although we were not always able to find hard evidence correlating the companies' external orientation to innovative success, the impressions we received from qualitative analysis indicate that front-runners tend to look outside the company more than pack members do. The differences in external orientation are not so much related to the outsourcing of R&D money, the attribution of more value to outside sources or even external co-operation. But differences in external orientation do seem to be related to internal aspects such as being prompt to adopt new technologies, being able to sense new technological developments at an early stage, taking a lead with cost reductions by implementing new process technologies, and being actively involved in technological knowledge transfer.

Furthermore, from the analysis it appears that process innovations by pack members are initiated more frequently through experience with the current process (with the production department and management team as the most important sources) than through developments outside the company. With front-runners, process innovations appear to be particularly initiated by the more innovation-orientated departments, R&D and engineering. These findings are supported by Jovanovic and MacDonald (1994), who found that technological leaders tended to rely on innovations to reduce their production costs, while the laggards rely more on imitation. From this, we could posit the new hypothesis that process innovations by pack members are usually less innovative than process innovations by front-runners.

New hypothesis
> **Process innovations by pack members are usually less innovative than process innovations by front-runners.**

Both front-runners and pack members say that the most important co-operative relationships with outside parties in respect of technological co-operation are with their suppliers. For the pack members, competitors are in second place. This includes both pre-competitive co-operation and co-operation following market introduction. For the front-runners, however, (semi-)public research institutes and universities have second place. Although these differences are not statistically significant, this could indicate that front-runners are usually more advanced than their competitors and thus have less to gain from pre-competitive co-operation. This links up with the finding that front-runners have a technological lead on pack members.

Absorptive capacity
The 'technological leadership' scale thus failed to reach the significance level for inclusion in the regression equation. This can be attributed mainly to the fact that this scale correlates closely with the 'technological absorptive capacity' scale ($r = 0.70$, $p < 0.001$). This correlation comes as no surprise, since it can be argued that technological absorptive capacity (Hypothesis 5.3b) and technological sophistication (Hypothesis 5.2a) will be closely related. For instance, Henderson and Clark (1990) show that organisations with a high absorptive capacity will tend to be proactive, constantly exploiting opportunities present in the environment. Furthermore, accumulating knowledge in one period permits more effective accumulation in the next. Basically, because of its technological knowledge, the company can better understand and thus evaluate the importance of advances in technological development. By investing internally, a company learns to cope more effectively with the knowledge available outside. But the causality also works the other way. A company that innovates at the leading edge of technological knowledge and whose technological progress becomes closely tied to advances in basic science will increase its basic research. Furthermore, Henderson and Clark argue that, as the environment becomes more diverse, companies will increase R&D as they develop absorptive capacity in each relevant field.

5.3.3 'Innovative' or 'successfully innovative'?

Both the companies as well as the sectoral experts indicate that front-runners are more innovative than pack members. But we have to differentiate between innovative and successfully innovative companies. As was discussed in subsection 5.1.2, an important difference (in terms of definition) between an innovating company and a successful innovating company is that the latter is also able to exploit its innovativeness successfully in commercial terms. Basically, then, one could say that it is a distinction between investments in innovation and the effect of these investments on the balance sheet. To be innovative, resources must of course first be put into the system. The important thing is to spend these resources effectively, so that the efforts will also generate commercial success. Having an ambitious corporate strategy with a clear sense of direction is an important factor in this respect. This is also the lesson taught by Hamel and Prahalad (1989): having a long-term goal forces the company to take a big step (stretch) and employ its resources and capabilities with the utmost effectiveness and efficiency (leverage). During the interviews, for instance, some experts pointed out companies which can be classified as innovative but which lack the power to turn this into sustainable success.[3] Some of these companies even went out of business. For the current population as well we see that innovativeness is not synonymous with innovative success or company profits. Hence, some technology-related aspects appear to correlate more closely with innovativeness than with innovative success (Figure 5.3). The relative size of R&D expenditures is closely correlated with both the innovativeness of the company (0.51) and its innovative success (0.44). The correlation with the latter, however, is lower than with the former, indicating that high R&D expenditures are a better predictor of innovativeness than of innovative success. This is underlined by the fact that the correlation between the relative size of the R&D ratio and the ability of the company to generate profits from new products is even lower (0.39). Furthermore, as Figure 5.3 shows, being prompt to adopt new technologies also correlates more closely with innovativeness than with innovative success. *Being creative* and *technologically driven* appear to correlate slightly significantly with innovativeness, but not at all with innovative success. Furthermore, *technological sophistication* is closely correlated with innovativeness, but not at all with innovative success.

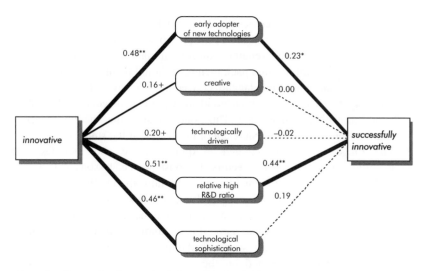

Note: Significance levels: + ≤ 0.10; * ≤ 0.05; ** ≤ 0.01

Figure 5.3 *Correlations between five technology-related variables and two*
dependent variables

Note, however, that this does not mean that these factors are unimportant for
the achievement of innovative success. They just do not differentiate pack
members from front-runners. This could indicate that an advanced
technological position in the sector, while frequently found in combination
with innovativeness, is not always accompanied by innovative success. An
alternative explanation, however, could be that the estimate of technological
sophistication is incorrect. After all, we are concerned here with self-rating
by the directors and R&D managers of the companies, and it could be that
the tendency to overestimate oneself is greater among pack members than
among front-runners. An indication of this is the fact that over half of the
pack members find that they are in the forefront of their sectors
technologically. We shall return to this point in greater detail in Chapter 9.

5.3.4 Methodological intermezzo

What has become clear from the analysis of technological competencies is
that the differences between front-runners and pack members are relatively
small. This came as no surprise and we shall see that it is also the case for
marketing and organisational competencies. Basically, three explanations can
be presented:

1. As a result of the research design, the gap between front-runners and pack members is small. As discussed in the previous chapters, we were looking for insights which can strengthen innovation management for the average company. This implies that pack members should be the contrast group to the front-runners and not laggards. A comparison with the latter would probably teach us more about good management in general than about good innovation management. Thus, one can state that the emergence of small differences is a logical result of the design of the study.

2. The range of sectors studied is quite diverse and includes both manufacturing and service sectors. In consequence, lessons will be at a higher level of abstraction and less unambiguous than with a single-sector study. This was also intentionally built into the research design.

3. The design is a cross-functional study with occasionally retrospective perspectives. For studies like these, it is common for the research subjects to have a tendency to regress towards the mean. This phenomenon occurs when subjects are selected on the basis of extreme scores on some variable (in this case, innovative success). When retested on the same or a similar variable (after some time), the original extreme sample generally tends to be less extreme (on average, scores will be closer to the mean). In our case, the front-runners category represents just such an extreme score and front-runners thus have a greater chance of becoming less successful than pack members. In this way, the differences become smaller.

NOTES

1. Customer involvement will be discussed in Chapter 6. It has been left out of this analysis because the focus in the current chapter is more on technologically related co-operation than on marketing-related co-operation.
2. For a scale to be acceptable, Cronbach's alpha should be ≥ 0.5 (Kaiser, 1974). However, an alpha of 0.5 is characterised as miserable and one of 0.6 as mediocre. An alpha of 0.7 is considered good (ibid.).
3. These companies were discussed during the expert interview but were not included in our sample.

6. Marketing competencies

> The problem is not that our customers want something we can't make, but that we can make things that our customers don't want. (Dr H. Gissel, Director R&D AEG, quoted in FEM, 1994)

6.1 EMPIRICAL FINDINGS FROM THE LITERATURE

New products and services are normally developed with the aim of putting them on the market and reaping the benefits. To achieve this, the new product should fulfil some apparent or latent customer need and be so attractive to customers that they decide to buy it. To this end, the company must assess the wishes and needs of its current, potential and future customers. And it should be able to translate these sometimes vaguely described ideas into new or improved products or services that satisfy the customer's needs. Such skills, which are aimed at understanding current and future user needs and translating these needs via concepts into products and services, are important building blocks of a company's marketing competencies, a construct we shall be devising in this chapter. Other important building blocks include the quality of communication with customers, promotion, and the ability to offer the product or service at the right place and through the right channels.

6.1.1 The importance of marketing for new product success

The critical role of marketing activities in achieving new product success has been documented by many studies. In the early 1970s, a number of US studies had already identified the importance of marketing variables for new product success. Inadequately done or mistakenly omitted marketing activities were shown to be a major source of new product failure (see, for example, Hopkins and Bailey, 1971; Cooper, 1979, 1975; Calantone and Cooper, 1979), inadequate market analysis prior to product development being the major culprit. Firms simply failed to do their marketing homework before they became seriously involved in the development of a product. The British SAPPHO studies showed that an understanding of customer needs,

154   Empirical findings

effective external communications and extensive marketing efforts contributed significantly to the successful completion of projects (Rothwell, 1972). Cooper's (1975) study of 114 new product failures also indicated that the major reasons for new product failure were chiefly related to marketing, with one-fifth of the cases overestimating the number of potential users of the product and almost one-fifth setting the product's price too high. In three-quarters of the failures, a detailed market study was either poorly done or omitted altogether. In contrast, technical and production activities were more proficiently and consistently undertaken. Although some weaknesses were identified in these areas as well, they occurred much less frequently than in the marketing-related activities.

These are all studies from the 1970s, but not much seems to have changed since then. One out of five firms in the recent 'Best Practices Study' by the Product Development and Management Association (Page, 1993) did not conduct preliminary market research to determine market need, niche, and attractiveness (concept testing), while concept screening (for example, scoring and ranking concepts according to certain criteria and eliminating unsuitable concepts) was not done by 24 per cent of the firms. Furthermore, the role of the marketing department in new product development is cited by one-fifth of the participating firms as an obstacle to successful new product development. The most recent NewProd investigation (Cooper and Kleinschmidt, 1993b) underlined earlier findings (Cooper, 1980; Cooper and Kleinschmidt, 1987a, 1987b) by stressing adequate marketing performance as a key determinant of new product success. According to Cooper and Kleinschmidt (1993b), key success factors for new products are 'delivering unique benefits to the customer' (product superiority), 'pre-development homework', 'sharp and early product definition', 'a strong market orientation with constant consumer contact and input', and 'quality of execution of key activities in new product process'. This study furthermore underlined some of the findings of the aforementioned 'Best Practices Study', notably with respect to the high number of firms which do not perform extensive market studies at the onset of product innovation trajectories. Cooper and Kleinschmidt found that such a study was not conducted in three out of four of the 203 new product innovations[1] studied. But even in instances where an extensive market study was undertaken, it was often done poorly. The quality of execution of market-orientated activities was shown to have a pronounced impact on success. Projects rated as having 'high quality marketing actions' were successful in 71.1 per cent of the cases, while only 32.5 per cent of the projects rated 'poor' for their marketing efforts were successful.

The picture in the service sector is comparable. In a study of 173 new financial services, Cooper et al. (1994) come up with basically similar conclusions. They identified market-driven new product processes as the dominant success factor for top-performing new service products.[2] Despite the vast amount of scientific and management articles on the importance of being market-driven, the researchers were surprised to see how weakly many projects scored on this aspect. Anecdotal evidence obtained during their research suggests that:

> most financial institutions have not learned the hard lessons that their counterparts who develop physical goods have: that a market orientation and market knowledge are keys to success; that a lack of market information is the number one reason for new product failure; and that most firms are sadly lacking in terms of their market orientation. (Cooper et al., 1994, p. 295).

But the service sector is catching up. Banks, for instance, are increasingly making use of market analysis, customer requests and focus groups to identify new products and services which can meet customer needs and improve quality. Some make prudent attempts to involve customers in product development, which should result in products that are easier to use, instil greater confidence, and add significant value to the customer (Teixeira and Ziskin, 1993).

6.1.2 Customisation

The ultimate goal of a company's marketing activities is to understand the customer's current and future wishes and needs, since such an understanding is an important prerequisite for new product success. Various studies have shown that 'understanding user need' is a highly significant discriminating factor between commercially successful industrial product innovations (for example, Rothwell, 1977; Freeman, 1982; Shaw, 1985) and those that fail. Similar results have been obtained regarding consumer products (Cooper, 1986). The importance of adapting to customer needs as an innovation success factor was also acknowledged by the sectoral experts who selected the companies for our study. Almost all experts regarded 'adapting products to customer demands' as the most important success criterion in their industry (Figure 6.1). Moreover, market-pull related criteria for innovative success was ranked 1–3.

Once customer needs are known, the challenge is to make products or deliver services which meet those needs. One important question in this respect concerns the specificity with which the market is divided into customer segments. At the extreme end we have segments containing only one customer. Adapting products or services to the specific needs of a customer or group of customers is referred to as *customisation*. The first time customisation was used as a competitive strategy was probably in the 1930s, when General Motors started designing cars for various life-styles and price categories. Before that, the car assortment was dominated by production requirements rather than customer needs. A famous and accurate characterisation of that era (the early 1920s) is provided by Henry Ford's immortal words: 'They can have any color car they want as long as it is black.'[3] In the 1960s and 1970s, booming industries and an exploding population more or less 'guaranteed' that many firms could sell just about anything they could produce. Product design was hardly an issue. The focus was on process innovations, process modifications and cutting production costs. Customers were interesting to the extent that they jointly formed the market. Ignoring the many differences, 'customer needs' were aggregated into 'markets'.

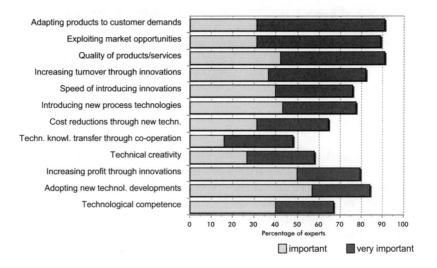

Figure 6.1   The relative importance of 12 measures for innovative success in 45 sectors (according to sectoral experts)

The turnaround came from Japan, where much more attention began to be paid to higher levels of quality; this increased the attention to the customers and their needs. The most striking examples of this were to be found in the automotive industry, which was increasingly allowing buyers to express their wishes for a 'custom-made' car. But other industries and services followed rapidly. Market segments were defined more and more narrowly to enable firms to respond more effectively to the specific needs of customers. This also had its effect on the management of innovation. While innovations in the 1960s and 1970s were usually generated by technology-push, market-pull became increasingly important in the 1980s. In the 1990s, it is by far the most important source of product innovations. A strong market-pull is not only a way to ensure that what is developed will be demanded by the market; it also pays off in terms of increased speed, as shown by a study of 103 new product projects in the chemical industry (Cooper and Kleinschmidt, 1994). Giving the customer a voice in the process by means of a customer-focused, market-orientated new product effort appeared to be an important key to success for on-time, fast-paced product development projects. In the 1990s, customer segments have become more fragmented than ever before. The reasons for this are numerous. For one thing, customers have become more demanding and more knowledgeable. Furthermore, suppliers have become more willing and capable (flexible organisations) to meet these fragmented needs. Nowadays, market segmentation is no longer sufficient for many industries. Even within segments, products have to be adapted to specific customer wishes. Customisation and customer-orientated development have become a major focus of innovative efforts at many firms. In the business press, the 1990s have even been labelled the 'Decade of the Customer' (*Business Week*, 1990).

Customer needs can be detected in many ways, from everyday, informal customer participation in a customer satisfaction survey or questionnaire about unmet customer needs, through the procedure for reacting to formal complaints, to the formal involvement of customers in innovation projects. Co-operation with the user or customer and even chain integration have become important elements in the innovation process for firms in industrial markets. Sometimes, the user plays a crucial role and becomes heavily involved in the manufacturer's development process. Users who experienced technical problems at an early stage and have already found (partial) solutions can play a crucial role in the manufacturer's innovation process. These 'lead users' (Von Hippel 1976, 1978 and 1982) turn to a producer of process equipment to get their ideas put into action.[4] Close co-operation with the customer allows the firm to detect new trends on the end markets at an early stage. In this manner, chain integration can have an 'early warning' function as well.

Most of the aforementioned studies were focused on the role of the customer in individual *projects*. The current study, however, is not focused on the individual project level but on the level of the *firm*. At that level, the debate focuses on the firm's common practices regarding customer involvement in general. However, it is to be expected that many of the lessons learned at project level can be translated to the organisational level. We therefore expect that successfully innovating firms will in general pay more attention to understanding customer needs than less successfully innovating firms, and hypothesise that:

Hypothesis 6.1

Successfully innovating companies are more customer-orientated than less successfully innovating companies.

We expect to find that front-runners, among other things, attribute more importance to market-pull sources for new product ideas and involve the customer in the innovation process more often than their less successful counterparts.

6.1.3 First-mover (dis)advantages?

Being the first in the market with a new product or service is expensive and risky. Sometimes, later entrants can leapfrog first-movers with superior products by incorporating into their products the lessons they have learned from the experiences of the innovator. The advantages for later entrants are obvious: the costs are lower and there is an example on which to base activities. One can, as it were, learn from the leader's mistakes.

On the other hand, the first-mover has the advantage of being able to build up a market share and set standards more rapidly. Moreover, in some sectors the image of being a technological leader plays a major role in buying decisions. Thus, being a first-mover is potentially very rewarding since advantages can be gained in terms of market share, customer loyalty, distribution, costs, and so on. The factors involved in achieving and sustaining a first-mover advantage are of course considerably more complex than a simple order-of-entry effect. Nevertheless, various conceptual and empirical studies have demonstrated causal relationships between first-mover status and long-term competitive advantages (Robinson and Fornell, 1985; Lambkin, 1988). In general, it is argued, first-movers develop sustainable market share advantages (Bond and Lean, 1977; Robinson, 1988) and are likely to be market leaders in their product categories.

Although these studies postulate a direct systematic relationship between order of product entry and market share, the line of causality is less clear. Are first-movers more successful than later entrants because they started with superior skills and resources? Or do firms gain a competitive advantage (and therefore superior skills) by being the first-mover? In the literature, two main streams of research on this subject can be identified (Robinson et al., 1992).

The first stream concludes that first-movers do not have a historical edge on later entrants, but that they develop their advantages by moving first (Urban et al., 1986; Lambkin, 1988; Robinson et al., 1992). In this view, the skills and resources of market pioneers differ from, but are not superior to, those of the later entrants. Moreover, resource requirements for success in a given business change with market evolution (Abell, 1978). The decision about when to enter the market, for example, is also influenced by chance and specific situational factors, such as the entrant's degree of product innovation, available distribution channels and expected competitive reactions. An empirical study by Robinson et al. (1992) supports earlier observations (Abell, 1978) that the strategic windows for market entry tend to open at different times for different entrant types. Kerin et al. (1992) have shown that product-market contingencies can help or hinder first-movers in taking advantage of their status. According to them, a multiplicity of controllable and uncontrollable forces determines to a large extent whether first-mover status leads to sustainable advantages.

Second, we can distinguish research based on the 'absolute pioneer advantage' hypothesis, of which Lieberman and Montgomery (1988) are the most prominent advocates. They posit that most firms basically want to pioneer new markets since market pioneering yields the highest economic profits, and that it is particularly the firms with superior skills and resources which will choose to pioneer new markets. Furthermore, they reason, one can assume that such firms (with superior skills and resources) will most likely be the winners in the contest. Therefore, market pioneers are intrinsically stronger than early followers and late entrants.

We tend to follow the reasoning put forward by the latter stream of research, and expect that the successfully innovating firms will most often be among the first to introduce new products, and that they most probably will be regarded as trend-setters in their own markets. We therefore hypothesise as follows:

Hypothesis 6.2
Successfully innovative firms have great influence on new product development in the markets they serve and are among the first to introduce new products.

In most of these studies, being a first-mover is defined as being among the first firms to sell a new product category. In our study, we did not expect to find many firms which had recently introduced a new product category. We therefore regard a company as a first-mover when it is among the first to sell a new or significantly improved product or service.

6.2   RESULTS

6.2.1   The role of the customer

Ways of acquiring knowledge about customer needs include asking
customers what they want, doing research on market developments and
monitoring competitors' activities. Companies frequently turn to these
market-related sources for ideas about new products. Other sources include
technology-related ones such as the R&D, engineering and production
departments, the literature, suppliers, basic research at universities, and so
on. We provided the marketing and research managers with a list of 18
possible sources and asked them to indicate the importance of each of these
sources for product innovations in the company. A factor analysis of these
sources resulted in seven factors. Next, scales were constructed from the
variables which loaded high on the respective factor. From Table 6.1, which
presents the T-test results, it becomes clear that none of these scales
discriminated significantly between pack members and front-runners. The
tentative conclusion is thus that front-runners do not differ from pack
members with respect to the importance they attribute to the various sources
of new product ideas.

As might be expected, Table 6.1 shows large differences in the value
attributed to each cluster as a source of ideas for product innovations. Fairly
unimportant sources include organisations doing research at the forefront of
technological change, such as basic research and public research institutes
(including universities). This is not very surprising as our sample consists of
low- and medium-tech companies. Upstream sources and business-related
sources rank as quite important, but by far the most important sources for
new product ideas are the downstream sources, sources close to the market,
such as customers, the marketing department and the unit management.
Thus, for front-runners and pack members alike, market-pull appears to be
more important as a source of new products ideas than technology-push.

When analysing the downstream-related sources in depth, it becomes
clear that both front-runners and pack members regard the marketing
department and the customer as the two most important sources of new
product ideas in the past five years (Table 6.2) and that they do not differ
with respect to the importance they attribute to each of these sources. The
reason why *management* loads heavily on the 'downstream source' factor is
probably due to the fact that our sample included many small and medium-
sized firms in which the general manager (often also the owner) or CEO is
involved hands-on with many facets of the firm, including customer relations
and marketing. In some cases, the owner/manager was even the driving force
behind most of the firm's innovations.

Table 6.1 Sources of product innovations

Variable/scale	Average all Firms	T-value front-runner versus pack member
Factor-derived scale [†‡]		
1. Downstream sources (customers; marketing; management of unit)	3.67	0.07
2. Business-related sources (competitors; corporate staff)	2.64	0.86
3. Upstream sources (R&D; suppliers of raw materials; suppliers of production installations)	2.60	0.15
4. External scientific sources (literature; symposia; congresses; courses)	2.58	0.16
5. Production-related sources (production; engineering; maintenance)	2.26	0.33
6. Public research institutes (universities; (semi-)public research institutes)	1.86	–0.17
7. Basic research (internally conducted basic research)	1.74	–0.61

Notes: [†] 1 = unimportant; 2 = fairly unimportant; 3 = important; 4 = very important; 5 = single most important source. [‡] The scales were derived from the variables which loaded high on the respective factors; Kaiser-Meyer-Olkin measure of sampling adequacy = 0.61

Table 6.2 Knowing the customer

	Pack members	Front-runners	T-value
Downstream sources of product innovations[¶]			
Customers	3.81	4.00	–0.91
Marketing department	3.83	3.69	0.57
Management of the organisational unit	3.38	3.29	0.39

Notes: [¶]1 = unimportant; 2 = fairly unimportant; 3 = important; 4 = very important; 5 = single most important source.

The problem, however, is that future customer or user needs can be quite elusive, since the users themselves have generally not formulated their needs consciously and are often unable to describe the products which could meet them. This is even more the case in consumer markets than in intermediate markets. It can be argued that even when users say that they are aware of their wishes and needs and are able to formulate them, there is a great risk that these needs will be closely linked to and heavily influenced by the current products and services. Firms then risk falling into the trap of 'new product myopia', only gradually improving on existing products and services.

In general, both front-runners as well as pack members often co-operate with customers with respect to innovation. The results indicate (though not significantly) that pack members appear to co-operate even more often with their customers than front-runners do (Table 6.3). Similar results were obtained when the respondents were asked more specifically about the stage at which customers are generally involved in one of the company's new product development processes. Neither the factor-derived variables nor the individual variables appear to correlate significantly with success. From this we can conclude that successfully innovating firms do not differ from less successfully innovating firms with respect to the frequency with which they co-operate with the various external parties. Most often, firms seek partners within the vertical supply chain. On average, such partners are quite frequently involved, and competitors and (semi-)public research institutes become involved only rarely.

Table 6.3 Involving the customer

	Pack members	*Front runners*	*T-value*
Customer involvement			
Co-operating with customers[†]	3.60	3.49	0.59
Involvement in idea-generating phase[‡]	3.69	3.47	0.82
Involvement during innovation process[‡]	3.48	3.44	0.15

Notes: [†] 1 = never; 2 = rarely; 3 = sometimes; 4 = often; 5 = always; [‡] 1 = not at all; 3 = roughly; 5 = intensively.

Other potential co-operative partners include competitors and suppliers. A factor analysis of seven possible partners outside the firm identified three main groups within the value-added chain: vertical, horizontal, and outside the chain (Table 6.4). The Kaiser-Meyer-Olkin measure of sampling adequacy is mediocre, but still acceptable (Kaiser, 1974).

Table 6.4 Co-operation with respect to innovation

Variable/Scale	Average pack member	Average front-runner	T-value
Factor-derived scales[‡]			
1. Not in value added chain (regional technology advisory centres, universities, (semi-)public agencies)	2.00	2.29	−1.57
2. Horizontal (pre-competitive co-operation; co-operation in the market with competitors)	2.13	1.98	0.82
3. Vertical (suppliers, customers)	3.47	3.49	−0.13

Notes: Kaiser-Meyer-Olkin measure of sampling adequacy = 0.55; * $\alpha \leq 0.05$; ** $\alpha \leq 0.01$; *** $\alpha \leq 0.001$ (one-tailed); [†] 1 = never; 2 = rarely; 3 = sometimes; 4 = often; 5 = always; [‡] The scales were derived from the variables which loaded high on the respective factors.

Neither the factor-derived variables nor the individual variables appear to correlate significantly with success. From this we can infer that successfully innovating firms do not differ from pack members with respect to the frequency with which they co-operate with various external parties in the innovation process. Both pack members and front-runners most frequently involve partners in the vertical supply chain. On average, these groups are involved quite often. Firms only rarely seek co-operation in the horizontal groups or outside the value-added chain.

Despite the fact that front-runners and pack members hardly differ with respect to the extent to which they involve customers and regard them as important sources of new product ideas, they do differ in the extent to which they regard themselves as market-driven. While, on average, pack members say that the term 'market-driven' is 'applicable' to their organisation, front-runners regard it as 'very well applicable'. This difference is highly significant ($t = -2.87, p < 0.01$).

If customer orientation is defined by the extent to which the customer or the marketing department is regarded as an important source of new product developments and the extent to which the customer is involved in the innovation process, we cannot conclude that front-runners are more customer-orientated than pack members. This means that Hypothesis 6.1 cannot be confirmed on the basis of our analysis.

6.2.2 Dominating the market

The evidence presented in Table 6.5 gives support for Hypothesis 6.2. Front-runners do differ from pack members in that they regard themselves more as trend-setters[5] than pack members do. Front-runners see themselves as being prompt to seize market opportunities.

Table 6.5 Knowing the market

Variable/scale	Average pack member	Average front-runner	T-value
Prompt to seize market opportunities†	3.55	4.03	−2.29**
Trend-setter‡	3.45	4.00	−2.36**
Turnover generated by largest product group (%)	64	55	1.55

Notes: * $\alpha \leq 0.05$; ** $\alpha \leq 0.01$; *** $\alpha \leq 0.001$ (one-tailed); † 1 = firm is laggard, 2 = firm is in tail of pack, 3 = firm is in middle of pack, 4 = firm leads pack; 5 = firm is front-runner; ‡ 1 = not at all, 2 = hardly, 3 = somewhat; 4 = quite well, 5 = very well... applicable to this firm.

We asked the marketing managers to list the five most important product groups in terms of turnover. We then asked them to indicate the market history for their largest product group in terms of introducing new products. The scale ranged from 1 (late follower) to 5 (market leader in the sense of being the first to introduce innovations). This variable is referred to as 'new product introduction lead time' and was determined for the largest product group of each organisation. Next, the extent to which this 'self-evaluation' could be supported by historical evidence was tested in an open discussion with the respondent. In a few cases, the respondent decided to change the score in the light of the discussion.

From the analysis, it becomes clear that front-runners are among the first to introduce new products. Almost 60 per cent of front-runners regard themselves as market leaders in their main product group, whereas only 32 per cent of pack members classify themselves as such. Although pack members run a good race too, they generally tend to regard themselves as members of a leader group (46 per cent of pack members), not as the leader[6]. The dominance of front-runners cannot be attributed to the fact that they might be more specialised in their main product group than pack members. On the contrary, pack members and front-runners show no significant difference in the portion of sales achieved by the firm's most important product group.

6.2.3 Contextual inspiration

When the hypotheses appeared to be relatively unproductive, we analysed the data to examine how front-runners did differ from pack members with respect to marketing competencies. As noted in Chapter 3, we are actually employing a 'trawler' method of data collection. Inspired by the literature, we began to look through the mountain of data produced in this manner for possible explanations of the difference between front-runners and pack members in the field of marketing.

Customer input and lead time (speed of innovation)
Based on the tendency of innovation processes to accelerate, we first examined whether front-runners were quicker to innovate and, if so, why that was. But front-runners did not prove to be 'faster' innovators than pack members; the average duration of innovation processes did not differ significantly between the two groups. The analysis showed, however, a relationship between the average duration of an innovation process in a company and the relative importance attributed by that company to downstream sources of new product ideas. Companies which attribute much value to production, marketing and customers as sources for new product ideas[7] appear to have shorter lead times than companies which attribute high value to basic research as a source of new product ideas. The more downstream the source for ideas, the higher the correlation between the importance the company attributes to it and the average duration of the innovation process. So a company which attributes relatively much importance to basic research as a source for new ideas tends to have long average lead times. This is shown graphically in Figure 6.2.

The relation between the lead time of innovation processes and a relatively high importance ascribed to downstream sources of innovation ideas is not surprising. It clearly corresponds to the findings of earlier studies (for example, Cooper and Kleinschmidt, 1994), which showed that a strong market-pull new product effort is an important key to on-time, fast-paced product development projects. The explanation might very well be that these market-pull innovations are by nature more clearly defined since there is a clear target or even a customer who has shown interest. These facts usually lead to a higher goal congruence among the people involved. This could be somewhat different with technology-push dominated firms, where it is still too often necessary to 'sell' the idea to downstream departments. Or, as the general manager of a front-runner in the plastics processing sector put it, 'The hardest sales are internal sales.' Furthermore, there are of course more uncertainties connected with technology-push innovations, which are also usually more complex innovations that thus demand more time.

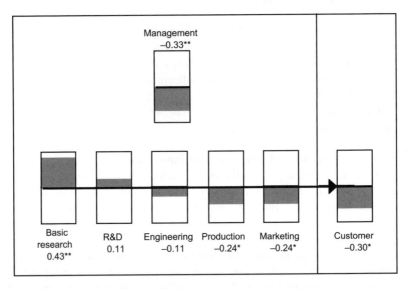

Notes: * $\alpha \leq 0.05$; ** $\alpha \leq 0.01$ (one-tailed)

Figure 6.2   Graphic representation of correlations between sources of innovation and average lengths of innovation processes

A partial explanation of why front-runners do not innovate faster than pack members might be simply because their innovations are more complex and/or more innovative. In that case, there would be little point in a general comparison of the duration of innovation processes between front-runners and pack members. It would be like comparing the ripening processes of apples and oranges. We still expect that comparable types of innovations will proceed more quickly with front-runners than with pack members, but this proposition cannot be investigated on the basis of the available data. A rapid average innovation time as such is thus not a characteristic of innovative success.

The firm's perception of its environment

Companies in each matched pair operate in the same market. One might expect that they would experience more or less the same environment.

Differences in the perception of that environment could therefore tell us something about their sensing capacity. We asked the respondents to indicate how they perceive their environment by means of eight scales (semantic differentials). The environment was defined as 'all external actors with which the company comes in contact and which influence the company's behaviour'. To ensure that each front-runner in the sample was paired with a pack member which experiences the same environment, the analysis was done on a subset which involved the 48 firms that formed 24 complete pairs.[8] A factor analysis on these criteria provided us with cognitive maps that indicate the way people think about their environment (Table 6.6).

Table 6.6 Factor analysis of environmental variables

	Factor 1 *Comfortable*	*Factor 2* *Dynamic*	*Factor 3* *Imposing*
Eigen value	2.3	1.5	1.1
Pearson's correlation with success (of scale derived from high-loading variables)	0.19	0.02	−0.10
Stable/changeable	−0.49	0.64[a]	−0.11
Predictable/unpredictable	−0.68[a]	0.15	0.18
Simple/complex	−0.26	0.61[a]	−0.03
Few/many customers	0.06	0.24	−0.66[a]
Hostile/friendly	0.80[a]	0.09	0.25
Fast/no fast reaction required	0.65[a]	−0.26	−0.24
Little/heavy external pressure	0.14	0.86[a]	−0.06
Much/little freedom	−0.03	0.09	0.87[a]

Notes: Kaiser-Meyer-Olkin measure of sampling adequacy = 0.64 (acceptable); [a] Denotes high factor loadings; $n = 48$ (24 complete pairs)

Basically, three factors emerged. One factor indicates how 'comfortable' the environment was perceived as being, with high-loading variables measuring the environment's predictability and friendliness and its 'not requiring fast reactions'. The rate of change of the environment, its complexity, external pressure and whether it requires fast reactions jointly form a cognitive environmental characterisation which can be labelled the 'dynamic' factor. The third factor has been characterised as 'imposing' as it includes variables which indicate the degrees of freedom a company has or experiences (having only a few customers for instance limits a firm's degree of freedom).

Front-runners do not differ substantially from pack members regarding their perception of the environment. The two groups do differ significantly, however, on one individual scale: their perception of the predictability of the environment. Front-runners tend to regard their environment as being more predictable than pack members do ($t = 2.23$, $p < 0.01$). In fact, front-runners tend to regard it as slightly predictable (4.21), whereas pack members regard it as fairly unpredictable (3.33).

Competitive weapons
Data were also obtained about the critical success factors of the firms' products and services (Table 6.7). The respondents were asked to rate the influence of several product or service characteristics on the success of their products or services. Both groups regarded high quality as their most important product-related competitive weapon. The two groups differed significantly with respect to price and depth of assortment. Pack members attribute a significantly higher value to price as a competitive weapon than front-runners do. Their product strategy can be summarized as 'Quality for a reasonable price', while front-runners appear to compete more on the basis of 'Quality and service'. For them, service is the second most important factor after quality. For pack members, service ranks fourth.

Table 6.7 Success characteristics of products and services

Variable†	Pack members	Front-runners	T-test
Quality	4.38	4.38	−0.02
Service	3.90	4.06	−1.00
Accurate delivery	4.03	3.88	0.92
Flexibility	4.02	3.78	1.03
Depth of assortment	3.10	3.52	−1.79*
Breadth of assortment	3.17	3.46	−1.24
Price	3.71	3.34	1.88*
Newness	3.00	3.31	−1.29

Notes: * ≤ 0.05; † 1 = no influence on success of products/ services, 2 = little influence, 3 = quite some influence, 4 = strong influence, 5 = determines success to a large extent.

6.2.4 Marketing competencies and innovative success

From the marketing-related variables, only one scale could be constructed which generated a sufficiently high Cronbach's alpha: the scale measuring customer involvement during the innovation process (Table 6.8). The alpha of the scale measuring market dominance was unacceptably low (0.5). Therefore, individual variables had to be included in the regression analysis of the marketing competencies as well. Circumstantial variables identified in the preceding section as possible indicators of marketing competencies were not used in this analysis as they themselves do not directly influence innovative success.

Table 6.8 Scales constructed from individual variables

Scale	Scales or variables used to construct the scale	Cronbach's alpha	Correlation with innovative success
Customer involvement		0.7	−0.07
	Co-operating with customers		
	Involvement of customers in idea-generating phase		
	Involvement of customers during innovation process		
Marketing dominance		0.5	−
	Prompt to seize market opportunities		
	Trend-setter		
	New product introduction lead time of largest product group		

Table 6.9 presents the results of a stepwise multiple regression analysis toward innovative success on marketing competencies. The regression analysis (which was corrected for multicolinearity) resulted in an equation with four variables: market drive; new product introduction lead time for largest product group; trend-setter; and prompt to seize market opportunities. From this we learn that 26 per cent of the variance in innovative success can be explained by the combined influence of these four indicators of marketing competencies. None of the variables in the equation proves to be significant, but this is due to the fact that the variables are highly correlated among each other. Therefore, little attention should be given to the individual betas.

Table 6.9   *Multiple regression analysis towards innovative success on*
              marketing competencies

Independent variable	B	Beta	t
Market drive	0.056	0.180	1.42
New product introduction lead time	0.055	0.175	1.46
Prompt to seize market opportunities	0.052	0.177	1.33
Trend-setter	0.061	0.202	1.47
(Constant)	0.624		21.28**

Notes: $R^2 = 0.26$, $F(4,56) = 4.87$**; * $\alpha \leq 0.05$; ** $\alpha \leq 0.01$

6.3   DISCUSSION: THE CUSTOMER AS KING, OR THE KING AS CUSTOMER?

The value of customer-orientated innovation has been preached by consultants, practitioners and theorists for years now, and one could say that firms have learned these marketing lessons well and seem to live up to them. The message of customer orientation has found its attentive listeners in all companies, front-runners and pack members alike. Both groups preach client-orientated innovation and the results show that front-runners hardly differ from pack members with respect to the importance they attribute to downstream sources of new product developments (the customer and the marketing department). Furthermore, both groups show hardly any difference in the extent to which they involve customers during the innovation processes. It appears that customer orientation is no longer a factor discriminating between front-runners and pack members. They claim it is something every competitive company has to do just to remain competitive.

So how can one stand out? Can a company gain a competitive advantage based on marketing knowledge and routines? Apparently it can since, despite the similarities, front-runners do seem to be able to differentiate themselves from pack members in terms of the extent to which they are recognised as customer-orientated innovators. They are more often regarded as trend-setters and innovative leaders than pack members. Furthermore, the results indicate that front-runners are quicker to exploit market opportunities than pack members and that front-runners regard themselves as more market-driven than pack members do. The question, then, is: What do front-runners do differently with respect to marketing competencies? In the previous section, it became clear that we cannot answer this question on the basis of the quantitative data we have gathered.

Qualitative data, however, do give us some clues to explain the difference in performance. These data do not represent hard evidence based on statistical correlations but should be regarded as new questions emerging from discussions within firms. The anecdotal evidence from our interviews indicates that front-runners go beyond the simple marketing lessons. They are more aware of the limits to the diversity of their assortment; they have a better insight into the vertical chain than the corresponding pack members and they pay more attention to service in connection with the physical product. Each of these aspects will be discussed in more detail in the following sections.

6.3.1 Insight into the vertical chain: the customer's customer

A number of explanations can be proffered for the apparent paradox in the previous section. For example, it is not only customer input which determines whether or not a firm will be a market leader. Furthermore, there appears to be a great difference between perception and action. In a manner similar to Argyris and Schön's (1978) contrast between espoused theories and theories-in-use, we have found that there is a discrepancy between how people think and how they act: the organisation is espousing one thing and doing another. Espoused theories represent the official organisational strategies, procedures and norms, while theories-in-use represent theories of action as they are used in the real world by the enterprise's individuals. So knowing what customers want is one thing; gearing innovative efforts to suit these needs is something else again. After all, market research only makes sense when it leads to concrete action in the company. Since it was not measured, we cannot make our perceptions in this respect quantitatively hard, but we obtained the impression that front-runners are more consistent in this than pack members. With front-runners, the results of customer surveys usually play a greater role in guiding innovative efforts than is the case with pack members. A front-runner which does regular research into customer value states that it is not unique in its sector for doing so. But, as the director pointed out, they have an advantage over their competitors in that they recognise customer value as a binding agent on which to focus future activities. Based on this observation, we can posit a new hypothesis:

New hypothesis
Front-runners use customer input and results from market analysis more thoroughly than pack members.

A possible explanation for the differences can also be sought in the manner with which the company deals with customer information. After all, listening to the customer has many gradations. It is important to ask what happens to this information. In this respect, there is a great difference between the practice in company 26 (front-runner) and company 43 (pack member). While company 26 seizes every opportunity to do something with customer information, company 43 allows opportunity after opportunity to pass by unused (Box 6.1).

BOX 6.1 JUST LISTENING IS NOT ENOUGH!

Front-runner 26 says it knows its customers very well. This company supplies fodder to farmers and, as an additional service, advises them on the use of its products and other related matters. Although the company looks on farmers as its customers, there is little direct contact between the sales department and the farmers. The products are sold to farmers via some 120 independent dealers. For this reason, the company has engaged a number of field workers (consultants, information officers) who visit the end users (farms), both to keep in touch with the market, and also to advise farmers on the use of the company's products. In order to make optimum use of the information they acquire from customers, the field workers write short reports on each customer contact. These reports are emailed to the head office and from there sent on to the various interested parties in the organisation. In this way, the information is distributed to places where some use can be made of it.

Company 43 (pack member) also listens to its customers, but much of this information becomes lost. For example, the company conducted an end-user survey in the past which generated a number of interesting things. However, the marketing manager admits openly that so far nothing has been done with this survey. At the moment, in fact, he is busy with an audit focused on the 'lost customer'. Discussions with the director and the R&D manager reveal that they do not expect much to come out of this study. The marketing manager appears to be a voice crying in the wilderness.

Furthermore, as Christopher et al. (1994) argue, 'In discussions with literally thousands of middle and senior managers it has become clear to us that only a relatively small percentage [...of the companies that claim to be marketing orientated, customer focused, customer orientated or marketing led...], perhaps 30 per cent, practise what they preach'. As was stated in subsection 6.1.1, previous research has shown that no or inadequate market analysis prior to product development is the major culprit blocking new product success (for example, Hopkins and Bailey, 1971; Rothwell, 1972; Cooper, 1979; Calantone and Cooper, 1979; Page, 1993). Many companies simply do not do their 'pre-development marketing homework'. Approaching 30 years after these studies, our findings still underline their conclusions. It is surprising how poorly market research is sometimes conducted. We visited diverse companies which did listen to customers but interpreted the information incorrectly or simply ignored it. Frequently, what one hears is not what one thinks one is hearing. We received the impression that pack members were usually sloppier in their market research than front-runners, as illustrated in Box 6.2.

BOX 6.2 THE ART OF ASKING QUESTIONS

Company 63 (pack member), which produces durable consumer articles, conducted extensive market research regarding the introduction of a new product. The results were very hopeful: the concept had certainly caught on with the target group. Yet the market introduction of the product turned into a major and costly flop. Subsequent analysis by the company revealed that no one, throughout the entire study, had thought to mention the price of the new product. When asked whether they would like to own such a product, a significant part of the target group answered *'Yes'*. If they had been asked whether they would be willing to purchase the product at the price the company hoped to get for it, the results would undoubtedly have been considerably more gloomy. This question, however, was never asked, and it took its revenge following the market introduction. The product was a flop.

We realised that such an example is not unique when, a few months later, we visited a pack member in the food industry which had made exactly the same mistake. This company (number 52) developed a frying and cooking margarine which was specifically intended for the preparation of chicken. It had a number of additives which ensured a delicious chicken gravy. The market tests were promising: both customers and experts were enthusiastic. After four months, however, the product was withdrawn from the market due to very disappointing sales results. The taste was great and the product did what it claimed, but consumers did not appear willing to buy a separate margarine just for cooking chicken. This question had not been asked in the market survey.

The following example illustrates how listening to customers can be done effectively. Company 54 (front-runner) is active as a contractor and process automation expert. Service and optimum response to the needs and problems of customers are of crucial importance to the company. A special structure has been introduced to address the future needs of customers. The personnel who contact customers have been divided into two levels: the chief mechanics, who have regular contact with customers regarding current projects, and the managers, who usually contact customers once a week. The personnel who work on the customer's premises are trained to interpret signals from customers regarding their needs for the future. Once a month there is a meeting with all managers in which the predominant view of the future and the specialties which could then play a role are discussed. These ideas are conveyed to the R&D manager who, in consultation with the director, determines whether they should be translated into projects. To eliminate any misunderstandings before a project is actually launched, the director will speak with the chief mechanics to find out what the original information was on which the ideas were based.

But there is another difference which is at least as important and has hardly been documented at all in the literature. Among the group of companies in our sample which do not sell their products or services directly in the consumer market, there appear to be front-runners who look further than their immediate customers in the vertical chain. They conduct marketing research not only in the immediate markets but also in markets further down the value-added chain. They sometimes even spend a substantial part of their advertising budget on their clients' clients. And these are not just the top-brand firms for whom it is normal to advertise to end-user markets. We have found front-runners which produce intermediate products but regard monitoring end markets as a factor crucial to their success. Front-runners often perform end-customer studies to get an insight into core customer values. They then focus their innovations on these core values. These firms have a clear picture not only of their direct users but also of the end-market customers. Pack members often lacked these insights (Box 6.3). These findings underline heavily the importance of the early-warning function of chain integration, as was discussed previously.

BOX 6.3 THE CUSTOMER'S CUSTOMER

The difference in insight into customers demands can be very effectively shown by the company pair in the furniture industry. The front-runner (company 56) restricts its attention to the distributive trade but none the less manages a number of showrooms to maintain contact with the 'end consumer'. According to the director, this is one reason why the firm has a very good idea of the needs of 'the consumer'.

The pack member (company 57), on the other hand, is poorly informed about the needs of the end consumer. To a significant extent, the management relies on the distributive trade and its own impressions. Until recently, as a matter of fact, the firm only bothered with direct buyers. Disappointing results forced the company to face facts. Looking back on past years in an interview, the director concluded: 'We actually listened only to the people who paid the bills, and that was our greatest mistake.'

Although the perception of the environment as such cannot be regarded as a marketing competency, it does give us some indications about the firm's sensing capacity, and this sensing capacity can be regarded as a marketing competency. Morgan (1992), for instance, states that in turbulent times organisations need to develop competencies that enable them to detect at an early stage the often subtle changes that are reshaping the contexts in which they operate and to cope with emerging threats and opportunities in a timely fashion. Since both firms forming a pair have more or less the same kinds of environment, differences in the perception of that environment can be regarded as signs of difference in knowledge about the environment. It is surprising that front-runners tend to regard their environment as being slightly predictable, while pack members regard it as slightly unpredictable ($t = 2.04$, $p < 0.05$), since the perception of both groups with respect to the changeability of the environment does not differ. They both regard their environment as equally dynamic ($t = 0.45$, $p > 0.30$). This gives us an indication of the difference in knowledge about the environment between the two groups. From this we might conclude that front-runners may have a better knowledge about their environment or at least believe that they have a better feeling for it. One could regard this as an indicator variable which says something about the firm's sensing capacity and knowledge of environmental changes. As has been discussed, qualitative analysis indicated that front-runners tended to look further down the value-added chain than pack members. Front-runners were much keener to monitor end markets and also focused their marketing efforts on these markets. Perhaps this knowledge about markets further down the value-added chain makes the environment more predictable. That may explain why front-runners find the environment reasonably predictable, while pack members experience the opposite. After all, if one keeps an eye on developments in the consumer market, one will have more insight into possible developments in one's own markets, which will make it easier to predict changes in the environment.

An alternative explanation might of course be that front-runners, who are often among the market leaders, have more influence on the environment, thus making it more predictable for them.

The fact that front-runners co-operate less with competitors following market introduction might also be an indicator of their possible market leadership. This reasoning is supported by the fact that front-runners are seen as trend-setters and that they usually manage to be first in putting new products on the market.

6.3.2   The customisation paradox: knowing when to say no!

Driven by a longing for customer orientation, companies sometimes tend to put a wide range of product variations on the market. The reason for this can be found in both the market and the technology. Thus, a highly market-dominated innovation strategy can lead to a broad spectrum of product innovations. But a strong technological position may also stimulate an expanding product line. Technologists have the tendency to innovate on the basis of the technologies they master. And if the possibilities offered by these technologies for product innovations correspond well to existing or latent customer needs, companies will be inclined to respond to these customer needs with product innovations. After all, one can make what the customer needs.

In section 6.2, it appeared that front-runners attach more importance than pack members to the 'depth of assortment' criterion. It is difficult to explain this observation. In all probability, we were not able to measure what we wanted to measure in this case. In fact, many front-runners attributed high value to 'depth of assortment' but added verbally that the assortment should not become too deep. Depth, in the sense of 'limited depth', was an important success criterion for them. The study also provided other indicators pointing to the fact that it is possible to go too far in meeting customer needs. It sounds paradoxical, but customer orientation can sometimes lead to a company's downfall. We refer to this as *the customisation paradox*. Apparently, customisation is not solely a route to success, but can become a route to failure as well. The latter will be the case if the internal variation becomes too great and the company has difficulty in managing all the complexity. Impelled by the drive to customise, the company will get bogged down with ideas for variations on existing products. If the organisation tackles too many such ideas, the innovation chain will become blocked with little projects costing considerable time and money. There will then be no time for innovative 'leaps' involving the next generation of a product or a completely new product. A number of pack members are now facing this situation (Box 6.4).

BOX 6.4 CUSTOMER ORIENTATION SHOULD NOT DEGENERATE INTO ZIGZAGGING

The history of company 37 (200 employees, manufacturing industry, pack member) has been strongly dominated by production and production technology. This dominance is clearly expressed in the organisational culture. Through mistakes in the past, the organisation has learned its marketing lessons. Currently, high flexibility with regard to customer needs is thus considered to be one of the most important success criteria. According to the director, this is a way of minimising the risk of selecting unsuitable projects. It also involves a challenge for the development department to realise these needs technologically. The customer demands are so diverse, however, that the internal organisation is becoming choked by all sorts of small modification projects. Moreover, reactions from the market are responded to in a very *ad hoc* manner. The recently appointed marketing manager puts it as follows: 'We are incapable of maintaining the same list of priorities for three months.' Each project is discussed in the management meeting. Even marginal projects, such as changing the product labelling for the Norwegian market, are on the agenda. For a long time, this company lacked insight into the costs of the innovation processes. A more detailed analysis of projects recently concluded by the marketing manager reveals that the company makes a reasonably large profit on the standard products, but suffers considerable losses on the customer-initiated versions.

The marketing managers in a number of companies talked about the large number of new product ideas they had come up with. Virtually without exception, they were all based on concrete customer needs. For this reason, companies were surprised that many of their new products failed on the market.

For company 46 (front-runner in the insurance business), innovation refers more to the adaptation and simplification of existing products than to constantly thinking up new products: 'For us, innovation means developing four new products (services) per year, with the goal of reducing the total service package from 100 to 20. Innovation is growing by 10–20 per cent per year with increasingly fewer personnel.'

According to the experts, this company was a laggard ten years ago, a pack member five years ago and is now definitely a front-runner.

The crucial challenge in this respect is to find a balance between the growing diversity of customer needs (external variation) and the inevitable need to simplify internal processes (internal variation). Front-runners appear to be more capable of finding this balance.

A considerable number of front-runners are aware of their possibilities and limitations when faced with the dilemma of 'standardisation versus customisation' (Box 6.5). Front-runners appear capable of drawing conclusions from the market more quickly and can thus more quickly choose between a technology-based and a market-based focus. These firms tend to set boundaries to the range of variations sooner than pack members.

BOX 6.5 COPING WITH THE CUSTOMISATION PARADOX

Strategies applied by firms to deal with the dilemma of standardisation versus customisation include the following:

Having the client pay for special orders and unique specifications
Pack members tended to view these as investments in knowledge development which would be recovered from repeated orders.

However, they failed to obtain such orders. The time and money invested in 'future turnover' were thus lost. Front-runners appeared to be much more businesslike in this respect: extras must be paid for.

Customer selectivity Front-runners frequently define their markets very consciously and will not easily deviate from them. For some front-runners (for example, companies 7, 12, 41, 46 and 62), the implication was that they would lose a certain client group. However, the loss in turnover was more than proportionally compensated for by an increase in efficiency. The front-runner in the plastics processing industry has a distinct boundary. This company has made its mark with mass products sold at prices up to approximately 100 guilders. This is its rationale. It thus defines its market as consisting of plastic domestic articles with selling prices up to 100 guilders, which is also the maximum selling price for all innovations. Company 62 (front-runner) does not operate in the higher market segment where products with higher added value can be sold. After all, its success was based on marketing skills which enabled it to operate very successfully in the present segment. It requires courage to be able to ignore the enticement of higher added values.

Niche strategy Front-runners, particularly the smaller firms, are usually niche players. They seldom face fierce competition in their own market segments. They distinguish themselves particularly through the unique services they provide. A producer of fodder not only supplied the fodder but also the know-how and assistance needed to set up and equip the stables and farm.

Assemble-to-order A number of firms tried to control product diversification by developing modular products and services. Used in different combinations, the assembled products could be sold as 'tailor-made'. The advantage is that the modules can be produced as standards. In this way, internal variety is limited while more external variety is made possible.[9]

Talking the client into a standard product Sales personnel frequently invest too little energy in finding out whether a standard product can also meet specific customer demands. Often, specific demands are not as 'hard' as they seem and the client can also be satisfied by using a little persuasiveness or by reformulating a

question. More generally, firms can achieve a competitive
advantage by influencing rather than simply responding to
consumer preferences.

The motto seems to be: 'Adapting to customer needs is important, but
know your limits. Know when you have to say "no" and say it' (Box 6.5).
The smart companies know their limits when it comes to managing
complexity. Co-operative efforts are complex, and the management costs are
accordingly high. It is only possible to deal with great external variation
when the internal processes have been organised as simply as possible
(reduction of internal variation) and the right market choices have been
made: 'This is our trade and we stick to it.' This goes farther than the well-
known adage 'Let the cobbler stick to his last', for it also implies a choice
regarding the type of shoes to be cobbled on the last and the type of
customers to whom they will be sold. The less clever companies lurch and
zigzag all over the market.

This reasoning goes further than the linear relationships between
customer orientation and innovative success as they are sometimes presented
in the literature. The literature suggests that companies will become more
successful in their innovative efforts to the extent that they become more
effective in listening to customers and translating their needs into new
products. The relationship seems to resemble a straight, diagonally ascending
line. Our research, however, indicates that the relationship between customer
orientation and innovative success can be better represented by an inverse U-
shape. Customer orientation contributes to innovative success only up to a
certain point. Beyond that point, it can even become a hindrance to success.
A schematic representation of this hypothesised relationship is depicted in
Figure 6.3.

The study has thus led to a new hypothesis:

New hypothesis
   **The relationship between customer-orientated innovation and
   innovative success has an inverse U-shape.**

We are not able to test this hypothesis statistically on the basis of our
current data. However, we believe it is a promising issue for further research.

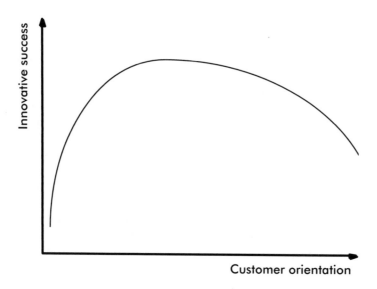

Figure 6.3 *The customisation paradox: a schematic representation of the hypothesised relationship between customer orientation and innovative success*

6.3.3 Problem solving as service

Front-runners appear to attribute significantly less value to 'price' as a competitive weapon for their products than pack members do. With quality ranking first, 'service' is the second most important competitive weapon for front-runners. The kind of service they offer goes beyond simple 'money-back guarantees', 'we-help-you-with-a-smile' policies, technical service or after-sales support. For many front-runners, service has evolved into a concept under which the entire company (processes, structures and culture) is organised around giving users what they want (expressed needs) or what the company believes the customers need (expressed as well as latent needs). The aim is to provide unique value in chosen markets, sustainable over time, which attracts new customers and brings existing customers back for more. Insights in how to achieve this can be found in the literature on relationship marketing (for example, Christopher et al., 1994). Relationship marketing has emerged as a concept to help realign quality improvement programmes, customer service initiatives and marketing efforts with the aim of increasing the effectiveness of the impact on the customer. The major challenge is primarily concerned with the establishment of enduring and mutually profitable relationships between the firm and its customers. Christopher et al.

(1994) maintain that relationships outside the organisation depend upon the quality of relationships within the organisation and that the strategic intent and shared internal values effectively become part of the products and services offered.

A number of front-runners have even included the process of thinking critically along with the customer as part of their corporate missions. And such words are put into action, as is illustrated in Box 6.6. One way in which these companies distinguish themselves from their competitors is through the extra service they provide in addition to their products. Their 'products' are thus more than physical products alone. With slogans such as 'Our products are solutions to customer problems', some front-runners present themselves very distinctly as 'problem solvers' rather than mere manufacturers of products. Others don't go so far, limiting themselves to providing advice and information. None the less, they manage to represent clear added value for their customers by doing so. Such companies attach great value to building up partner relationships with customers and taking their problems into account.

BOX 6.6 SERVICE CONNECTED WITH COMPETENCIES

Company 8 (front-runner in road transport) takes care of transport for its clients. It describes its product as the solution to the customer's logistical problems. The company is thus engaged not only for its 'wheels' but also for its comprehensive transport organisation. For company 8, the product is service: 'In principle, even the trucks are superfluous. We could easily rent them.'

Company 7 (front-runner in the electrotechnical industry) assembles regulating, control and measuring equipment for its clients. In the 1980s, it developed from a jobber (assembling printed-circuit boards in large batches against low prices) into a sparring partner for the client. The success of the products is based on the total package which is offered. Systems are developed and tested in collaboration with the client, with any problems which may arise being solved in the process. Contact with customers is thus of essential importance. The product is of high quality and competitively priced. The company describes its activities as 'solving customer problems'. As key words for its innovation policy, the director mentions 'assertiveness, creativity, daring and perseverance'. Assertiveness, as he sees it, is primarily to be found in the critical examination of customers and the specifications they put forward. This generally leads to changes in the specifications, which are then discussed thoroughly with the customer. In one case, the firm even turned down the assignment because the client did not agree with modifications to the specifications. For company 7, fear of harming its quality image was often reason enough to turn down an assignment. The pack member (company 15) describes its activities as 'assembling printed circuit boards on customer specifications'.

Company 26 (front-runner) produces compound feeders for pigs, poultry and cattle. The company defines innovation as 'everything which can lead to improvements for the customers, in both financial/economic and practical terms'. A significant piece of added value offered by company 26 to its clients is thus the provision of advice and information regarding aids to feeding the animals. Such aids vary from stable construction and furnishing to the manner in which the food should be distributed. An important part of the firm's sales is generated by means of this consultative function. In the past, innovations were primarily to be found in the area of service. Today, the accent has shifted to the product as it appears on the market, that is, including service. The provision of advice and information is also a way of finding out more about the end customer. In fact, knowledge about customers is primarily built up by the consultants who advise the farmers. In this way, the company is less dependent on dealers for information about customers.

Some companies which supply retail stores advise their clients (which are sometimes large national chains) how they can best bring their assortments to the attention of customers. A front-runner which produces household articles (company 62) has a couple of pages in each advertising brochure put out by its two largest clients on which it can plan the layout itself.

For front-runner 25, integration of services in its product line through collaboration with customers (chain stores) is becoming more and more important. Increasing use is being made of relation management in connection with clients. Thus, it provides the chain stores with advice about approaching consumers based on its own research into consumer eating habits.

Front-runners sometimes carry out cost-benefit analyses for their clients to show the value of their extra services in terms of savings to the customer or greater efficiency. In this way, the front-runner can make it clear that the relatively high price of its products or services is more than compensated for by savings to the customer (Box 6.7).

BOX 6.7 CONVINCING THE CUSTOMER

Company 62 (front-runner) manufactures a top brand of household articles and concentrates its promotional efforts very intensively on the end consumer. Its direct customers, however, are national and international chain stores. These clients are of course approached in an entirely different manner. Recently, company 62 found another new way to form closer bonds with its customers. When it realised that its close-knit, flexible distribution network was one of its most important strengths, it looked for ways to further expand this network. After some study, it was able to convince its largest client (a national chain store) that it could take care of the logistical planning for this chain's assortment more efficiently and economically than was possible with the distribution strategy employed by the chain up to that time. Based on a detailed cost-benefit analysis of the two distribution methods, the chain store was eventually won round and now pays company 62 for this extra service.

Due to a business dispute, a supermarket chain replaced a large part of company 25's assortment (food products in glass, front-runner) by a different brand. In reaction, the company had a study carried out with this supermarket chain into the market share of vegetables in cans or glass. It was thus able to convince the chain on the basis of hard figures that brand loyalty for its product was so strong that customers would prefer to shop elsewhere rather than make do with a second-rate brand. Since these customers would also shop for other items elsewhere, the supermarket chain was facing a considerable loss of sales. It wasn't long before company 25's products were again displayed on the shelves.

The increasing entwining of manufacturing production and service is a trend which has been found in many sectors, albeit to different degrees. Even some of the best-performing pack members show developments in this area. This development has significant organisational consequences. Products and services are increasingly provided via a single channel and the internal chain is becoming more highly integrated. One could almost speak of a process of evolution which companies must go through if they want to innovate in a more client-orientated manner, a process with the following phases:

Phase 1: Divergence   Companies which decide to innovate in a more customer-orientated manner (frequently by going in the direction of specialties) first go through a divergence phase in which many new products and variations are developed and put on the market.

Phase 2: Convergence   After some time, companies learn from their experiences in the market and the assortment is circumscribed and streamlined. The company now decides on a definite strategy which gives direction to further innovations.

Phase 3: Core product plus service   In this phase, companies start to distinguish themselves from their competitors through the service they provide in connection with core products. Some even manage to make themselves virtually indispensable to their customers.

The above phase model is a new hypothesis which could be posited on the basis of the results. The validation of this model will require further research. In this respect, we can remark already that the phases may also take place in parallel. A few companies have both products with integrated service packages and products without such packages.

Until now, customisation has referred particularly to the process of satisfying customer needs in terms of product specifications, order size, delivery times and quality. Firms which were quick to follow this direction are now carefully retracing their steps. Customisation is now translated into the advantages which the client can acquire by using the firm's solutions. In that sense, customisation is not only a process of listening to but also of convincing the client. What customers really need is as important as or even more important than what they demand. The supplier becomes the problem solver. Thus, the question as to where adaptation to clients ends and their education begins is of current interest. Not only does this imply a higher level of service, it also means being aware of and focusing on the needs of the customer's customers. Often, it also implies closer co-operation between the supplier and the customer or user. The question as to how far one can go in adapting to user needs still remains, however. What is technologically feasible? And when does a healthy diversification strategy become an unhealthy dance to the customer's changing tunes? Answering such questions demands accurate insight into factors which cause internal complexity and factors which obscure the view of broader market tendencies externally.

NOTES

1. Of which 123 were commercially successful.
2. A recurring theme in their research results was the need for more customer input from idea to launch, the key role of market research, market knowledge and competitive analysis, and the pivotal nature of a well-planned and executed new product process as important aspects in explaining the success of these innovations.
3. The reason was that black paint dried more quickly than other colours.
4. Although most of Von Hippel's studies on user involvement were conducted in intermediate- and high-tech industries, a study by Mantel and Meredith (1986) on technically complex, fundamentally generic and advanced innovations shows that Von Hippel's findings hold true in a diversity of firms.
5. The classification 'trend-setter' appears to be highly correlated with the ability of a firm to generate increased turnover as a result of its product innovations ($r = 0.59$, $p < 0.01$) and the overall innovativeness of that firm ($r = 0.55$, $p < 0.01$).
6. Pearson chi-Square of cross tabs: 8.28 (significance < 0.05).
7. No significant relationships have been found between the extent to which clients are involved in innovation processes and the average duration of innovation processes.
8. An analysis on all 63 firms revealed similar results.
9. For more information about the interaction between customisation and production systems (or the customisation–responsiveness squeeze), we refer the reader to McCutcheon et al. (1994). According to their framework, firms facing a low demand for customisation should employ a make-to-stock approach. The most appropriate strategy for high-demand customisation depends on the responsiveness of the customer and the position of the product differentiation stage in the production process. If customers are willing to wait for products, it is best to adopt a make-to-order approach. If customers demand quick delivery, a firm should opt for assemble-to-order (when products are differentiated in the final stages of the production process) or build-to-forecast (when products are differentiated at initial stages of the production process).

7. Organisational competencies

We must learn to use our intelligence to create simplicity rather than complexity. (General manager of a front-runner)

7.1 THEORY

Innovation is a process which has to be organised: 'Businesses must adapt to changing customer and strategic needs by establishing internal structures and processes that influence their members to create organisation-specific competencies' (Ulrich and Lake, 1990). To a great extent, the way in which this process is organised determines the level of success at the end of the chain. Organising and managing the innovation process is, in fact, a competency itself. We call it the organisational competency. Organisational competencies represent the firm's ability to manage people in order to gain a competitive advantage (ibid.) by means of its structure, planning procedures, culture or the way its innovative efforts are organised. They form the bases for activities that enable the company to be organic, adaptive and flexible. Companies with strong organisational competencies have created a working environment (routines and practices) that stimulates co-operative and risk-taking behaviour. This chapter discusses the types of structures, procedures and cultural characteristics that can be said to stimulate the successful management of innovations.

Lammers (1986) argues that the current views and theories on managing innovation have their roots in a study of organisational change and technology by Burns and Stalker (1961) that is already over 35 years old (Burns and Stalker, 1961). Based on a study of various organisational types in 20 companies in the UK, they argue that some organisational types are more suitable for innovative organisations than others. Burns and Stalker described two organisational types (the mechanistic and the organic) which are generally regarded (Mintzberg, 1983; Moss Kanter 1988; Hertog et al., 1996) as two extremes of a continuum along which most organisations can be placed: 'two polar extremities of the forms which [such] systems can take when they are adapted to a specific rate of technical and commercial change' (Burns and Stalker, 1961, p. 119).

The mechanistic management system is appropriate to stable conditions and can be characterised (ibid., p. 120) by:[1]

- functional concentration;
- centralised decision processes;
- hierarchic structure of control, authority and communication;
- individual responsibilities;
- top-down with predominantly vertical co-ordination and communication;
- an inward-looking mentality.

Organic structures, on the other hand, are more appropriate (ibid., p. 121) to changing conditions which constantly give rise to fresh problems and unforeseen requirements for action which cannot be broken down or distributed automatically to the functional roles. These structures, which function extremely well in the dynamic and complex environments of innovations, are characterised by:

- frequent lateral as well as vertical interaction;
- decentralised decision processes;
- a network structure of control, authority and communication;
- collective responsibilities;
- co-ordination by consultation;
- flexibility.

In this chapter, I show how front-runners and pack members can be placed with respect to this dimension. I do this at two levels: the level of the organisation as a whole and the project level. The following subsections discuss the two extremes of the continuum as well as several hybrid structures.

7.1.1 Mechanistic structures

Functional organisation
Mechanistic organisational structures are characterised by rigidity, regulated co-ordination, individual responsibilities and a distinct hierarchy of control. Burns and Stalker (1961) regard Weber's 'bureaucracy' (Weber, 1947) as an ideal type of a mechanistic structure:

> Bureaucracy, stands as the 'formal organisation' of industrial concerns... These general principles [of Weber's bureaucratic organisation] underlie every subsequent definition given to formal organisation in industry. (Burns and Stalker, 1961, pp. 107–108)

Building on classics such as Lawrence and Lorsch (1969) and Burns and Stalker (1961), Mintzberg (1983) elaborates on their ideas and presents five configurations.

Mechanistic structures are organised on the basis of functional concentration, an organisational order in which similar activities, skills and expertise are concentrated within the formal boundaries of the organisational structure. Functional concentration has two roots (Hertog et al., 1990): the separation of functions and the separation of 'thinking' and 'doing'. The first originates from the development of craftsmanship (Sitter, 1989). Production techniques have to be developed, learned and transferred. The craft organisation seems to provide the right environment (Mintzberg, 1979). Functional concentration with an emphasis on professions and functions is much in line with Weber's thoughts. These ideas can be traced in Mintzberg's 'professional bureaucracy' and Moss Kanter's critiques on segmentation as well. The second root stems from scientific management, which argues that the differentiation of activities into those with long and short learning times is expected to result in increased efficiency and that, through the development of routines, functional concentration will eventually result in increased speed and time savings. This represents another distinction: the separation between thinking (professions, staff, functions) and doing (production). We also encounter this distinction in Mintzberg's machine bureaucracy.

The functional approach is the traditional approach to organising product development. In functionally organised companies, projects are broken down into segments which are then assigned to the respective functional managers for co-ordination. Departments are often involved sequentially with the project. Within each department, large tasks are subdivided into smaller tasks, which subsequently must fit together again properly. As far as possible, activities are planned in advance. Considerable co-ordination is required to fit all the pieces back together again properly. The fact that the project has been subdivided into so many small pieces which must be allocated to various (functional) units makes a good deal of co-ordination necessary. In this model, co-ordination is effected through the hierarchy. People generally work on and support a number of projects simultaneously. The department head holds both line and technical responsibility for the project staff of his department. The overall co-ordination of the project is in the hands of functional management or higher managerial layers. Co-ordination is often achieved within a steering committee. Moss Kanter (1983, p. 28) calls this style 'segmentalism', as it is concerned with compartmentalising actions, events and problems and keeping each piece isolated from the others. Segmentalism assumes that problems can be solved when they are carved into pieces which are assigned to specialists working in isolation. In such a culture and structure, even innovation itself can become a specialty, something given to the R&D department to take care of so that no one else has to worry about it. Advantages of this approach are, among others, that a minimum of adaptations is needed in the organisational structure and that the work is usually of high quality, since specialists who readily share knowledge are grouped together. Furthermore, these kinds of organisational structures ensure great clarity in relationships and allow for flexible use of personnel. The functional approach is especially suitable for situations in which technical depth and excellent technical solutions are required and speed is not so important.

In the 1960s, however, when significant growth and expansion occurred among technology-based industries and the developmental aspect of R&D was becoming increasingly important and complex (Klimstra and Potts, 1988), the limits of a purely functional organisation in innovation management became clear. Increasingly, attention had to be paid to reducing costs, keeping deadlines and monitoring the progress of projects. But mechanistic companies experienced large problems in bringing about change. Companies which had to operate in complex and fast-changing environments with technologies and processes which were becoming ever more complicated and integrated, experienced difficulties in completing projects successfully by assigning, activities to different functional departments. Co-ordination became a major burden to them. Since innovations usually call upon the expertise of a wide variety of disciplines, stringent demands were imposed regarding the fine-tuning between functional departments. Increasingly, the disadvantages of these structures became a hindrance to effective innovation management.

Since these structures are input-orientated, both the internal as well as the external 'customer focus' is quite weak, often resulting in 'passing-the-buck' behaviour. And since functional success is generally not project-orientated, priority, co-ordination and accountability conflicts can be numerous. Burns and Stalker (1961) point to short-circuiting problems in communication, role conflicts within committees, 'superordinate' officials who report directly to the boss and interdepartmental conflicts. They argue that a rigid hierarchical structure is appropriate for a stable business environment, but may hinder organisational performance when the business environment is volatile (in which case a fluid, adaptable structure is proposed as more appropriate). In fact, mechanistic structures in general are adapted to relatively stable conditions.

Some of the main reasons for the inability of functionally organised companies to master innovation properly have to do with the fact (Hertog et al., 1990) that product development has a far more temporary character than production. The process is more complex, greater skills are required and the system's boundaries are constantly changing. Especially in science-based industries (pharmaceuticals, chemicals, electronics), new product developments often follow hard-to-predict trajectories in which the participants seldom know exactly what will come out in the end. Uncertainty reveals itself both in the resources (process, and so on) and in the objectives (what you are going to make and why). Takeuchi and Nonaka (1986) compare the traditional product development process with a relay race: a sequence of clearly defined development steps in which the baton is handed over to a new group of specialists at the end of each step. They point to the danger of batons falling to the ground. This is where most of the problems come to the forefront. Poor co-ordination frequently reveals itself in long development times, competency squabbles and products which lead to major delays in later phases (market introduction, production). In a functionally organised structure, the large number of departments involved in an innovation process will lead to external control problems (control being defined as the process of harmonising the subdivided process phases; Sitter, 1989), which will become larger if more departments become involved and the innovation is more radical.

For companies in more volatile business environments, a more organic structure (Burns and Stalker's other extreme) will be appropriate. Such structures are adapted to dynamic, rapidly changing environments in which new and unfamiliar problems are continually arising which cannot be broken down and distributed among the existing specialist roles. Organic structures are freer, more flexible structures in which decision-making is decentralised and responsibilities are shared collectively.

7.1.2 Hybrid structures

Functional matrix
Burns and Stalker (1961) argue that an essential part of a top manager's job is to interpret correctly the external uncertainties facing the company and decide on the appropriate management structure. The match between organisational structure and environment was further developed by Lawrence and Lorsch (1969), who introduced 'integration' and 'differentiation' as two opposing yet complementary concepts in constant creative tension with each other. In a study of organisations in three different industries, Lawrence and Lorsch (1967) found that the more varied and uncertain the environment of an organisation was, the more differentiated its structure needed to be (that is, into specialised departments able to cope with turbulent environments and technologies). But at the same time, the organisation has to ensure that all efforts are integrated; it must work as a single entity. Thus, the more differentiated the structure, the more necessary it is to put effort into the

integration of the various subunits. Their conclusion was that innovations call for structures which enable frequent interaction between departments and substructures and in which interdisciplinary co-operation and teamwork are stimulated. Matrix structures allow for meeting both needs.

In the late 1960s and the 1970s, then, many R&D-intensive companies responded to the aforementioned internal and external pressures to acquire more control of innovation processes by introducing formal project-control methods into the functional organisation. Firms began to resort to hybrid forms. One of the first attacks on the paradigm of functional organisations was provided by Marquis and Straight (1965), who stated that a functional matrix is likely to lead to better technical results than a functional organisation. In the functional matrix structure, a project manager is appointed to co-ordinate the project through the various departments. His or her authority is limited, however; the functional managers retain responsibility for and authority over their specific project segments. Functional lines continue to exist but receive an overlay of projects co-ordinating the activities of the various departments required to develop the innovative product.

The debate on matrix structures was opened. Fuelled by the increasing attention given to matrix structures in the behavioural sciences, the matrix as an organisational structure was gradually becoming a legitimate alternative for managers who wanted R&D to remain 'centralised' but were committed to the timely development of products (Klimstra and Potts, 1988). In the beginning, matrices only existed within R&D departments, but gradually other departments became involved with the matrix structure as well.

Balanced matrix

The increased power and influence of the project manager in the functional matrix eventually led to matrix structures in which responsibilities and power are equally divided between the functional managers and the project manager. The idea was simply that the functional line organisation would continue to exist, with the addition of an overlay of projects that would co-ordinate the activities of the various departments required to develop a particular product or service (Klimstra and Potts, 1988). In such a balanced matrix structure, a project manager is appointed to monitor and co-ordinate the project. Project team members are allocated part-time or full-time to the project, but continue to report to and operate from their functional departments. Technical responsibility for each function lies with the department head, while the project manager has overall line responsibility for the project and the day-to-day management of project team members.

To a certain extent, the matrix structure combines the advantages of a project team approach and a functional organisation. On the one hand, it allows a holistic approach, project emphasis and output orientation, while on the other hand it secures access to the pool of talent and maintains technical strength. Other advantages are that it integrates technical and business specialties. Disadvantages are mainly associated with the inherent dual reporting relationships, which can lead to decision-making conflicts, conflicts for resources between project managers, and a complex division of authority and responsibility. Furthermore, matrix structures require high commitment and energy to keep the organisation running smoothly.

7.1.3 Organic structures

Project matrix
In order to overcome some of the disadvantages of the balanced matrix approach, and in their efforts to speed up innovation processes even more, companies are increasingly placing greater emphasis on project management. Project managers have been given more responsibilities and two new structural arrangements have come into vogue: the project matrix and the project team. When a project manager has the primary responsibility for completing the project and 'borrows' team members or expertise from functional managers, the influence and importance of the project manager exceed that of the functional manager. This kind of structure is often referred to as a project matrix. While technical quality may be lower compared to a line matrix, the project matrix can enhance project integration and speed (Larson and Gobeli, 1985).[2] Increasingly, adequate interdepartmental and lateral relationships and the degree of cross-functional co-operation achieved have become important conditions for the successful completion of innovation projects (Pinto, 1989). Departments which are further downstream in the innovation process are increasingly seen as internal clients. In this connection, Bertsch and Stam (1990) refer to a 'market-in' approach (in contrast to the classic 'product-out' approach) in which the needs of the client come first and parties involved in subsequent phases are also considered as clients. In such an environment, problems are not distributed but shared. This approach leads to a culture which is conducive to more rapid problem-solving (Moss Kanter, 1983).

Project teams
The next step in decreasing the influence of the functional managers in the innovation process is to move to project team structures. Especially with respect to innovation, project team structures raised high expectations. Research has shown that teams produce more creative solutions (Osborn, 1957), make better decisions (Davis, 1973), improve the implementation of decisions and increase commitment (Cohen and Ledford, 1991; Hoffman,

1979). Furthermore, it has been indicated that certain tasks require teams because they can only be accomplished through constant mutual adjustment of the information provided by each team member (Donnellon, 1993). Thus, project-based and team-based structures came into vogue in many companies. Teams in which the various disciplines are represented are created to manage the project from beginning to end. Full-time team members are seconded to a project and sole responsibility for a project is given to the project manager, who often has total line responsibility for costs, time and product targets. Project teams imply horizontal rather than vertical organisation. The job of the functional departments is simply to supply capacity. Functional managers have no formal involvement in the project; the team is headed by a project manager. The 'full project organisation' structure enables strong customer focus (Table 7.1) and R&D/business integration. Communication lines are short and motivation is generally high. There are some drawbacks, though (Klimstra and Potts, 1988). Resource flexibility is generally low and technological (functional) expertise could leak away if the functional knowledge is not kept up to the mark. Instead of working like runners in a relay race (Takeuchi and Nonaka, 1986), the goal becomes to work more like a rugby team, with the whole team playing the ball around the field and moving back and forth together; with overlapping rather than sequential phases.

Some of the most striking differences between functional organisational structures, balanced matrices and a project team approach are summarised in Table 7.1.

The project team can be linked to the organisation at business unit, divisional or corporate level. The latter is often aimed at new businesses development, and is sometimes referred to as 'skunkworks' (Quinn, 1988b). These small, flexible teams often operate in a relatively independent environment. The concept originated at Lockheed, where small teams of engineers, technicians and designers were put together to develop a new product or process from concept to commercial prototype stages with no intervening organisational barriers (Quinn, 1988b, p. 130). Skunkworks and rugby teams are often cited as practices predominantly found in large companies, mostly in high-tech industries; such as Fuji-Xerox, Honda, Canon, Nec and Epson (Takeuchi and Nonaka, 1988) and Pilkington Brothers, Elf Aquitaine and Sony (Quinn, 1988b). In fact, such structures can be regarded as attempts by large companies to create an entrepreneurial spirit within the organisation, which is often regarded as a crucial factor for innovative success in small companies. An integrative view enables organisations to look at problems in their wider context and develop products which can be produced in the plant and distributed in the market (Moss Kanter, 1983).

Table 7.1 Characteristics of functional structures, balanced matrices and project teams

	Functional organisation	*Balanced matrix*	*Project team*
Focused on	Input	Output	Output
Resource efficiency	Medium	High	Medium
Resource flexibility	Medium	High	Low
Essential information flow	Medium	High	Medium
Clarity of relationships	High	Low	Medium
R&D/business integration	Weak	Moderate	Strong
Customer focus	Weak	Moderate	Strong
Unity of command	Not for project	No	Yes
Motivation/ identity	Low/Medium	Medium	High
Co-ordination/ accountability	Low	Low	High
Project manager	Lightweight (if any)	Lightweight	Heavyweight

Source: Adapted from: Klimstra and Potts (1988, pp. 45–46) and Roussel et al. (1991, p 141).

7.1.4 The full spectrum

Although the previous subsection considers the dominant streams in academic and practical thinking about organisations chronologically, all the organisational forms discussed can still be found in today's business community. Basically, they can be seen as a spectrum of organisational options which a company can choose to manage its innovation processes. The spectrum ranges from the functional organisational structures, in which full project responsibility is given to functional managers, through various matrix structures to the project team structures, in which full project responsibility is in the hands of the project manager. The resulting five basic organisational models are defined (Larson and Gobeli, 1988, 1985; Galbraith, 1971) as the functional organisation, the functional matrix, the balanced matrix, the project matrix and the project team (Figure 7.1).[3]

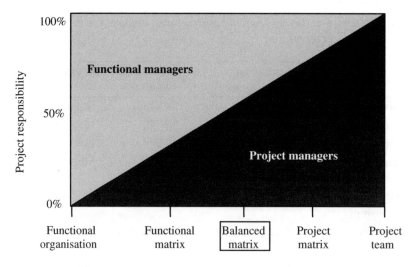

Figure 7.1 Five basic ways of organising innovation projects

Recent research points in the direction of project teams or matrix structures as the most preferred structures for achieving innovative success. Peters and Waterman (1982), for instance, argue that *ad hoc* project teams are the best approach for developing new services and products. However, they take a negative stance towards any form of matrix structures for product development. According to them, full or hybrid matrix approaches often become overly bureaucratic and uncreative, have a tendency to dilute priorities and can eventually degenerate into anarchy. Their evidence, however, is mainly anecdotal. More solid empirical evidence is provided by Larson and Gobeli (1985), who state that the balanced matrix is one of the preferred structures as it is likely to be flexible and adaptive. Since functional and project counterparts must jointly resolve disagreements, more balanced decisions can result. Similar results were obtained from a study of 86 R&D projects undertaken by Katz and Allen (1985). They showed that the highest performance was achieved when organisational influence was centred in the project manager and influence regarding technical details of the work was centred in the functional manager; that is, a balanced matrix in which the project manager is responsible for managing the project and the functional manager is responsible for the technical details. Nevertheless, it has been shown (Klimstra and Potts, 1988) that balanced matrices often lead to priority conflicts, power struggles and even inertia. In a later study, Gobeli and Larson (1986) report that managers rated the project team and the project matrix as the most effective structures for innovation processes. These results underline earlier findings by Corey and Starr (1971), who stress the importance of strong project management.

A study which looked at the whole spectrum from functional organisation to project team was conducted by Larson and Gobeli (1988). Based on a comparison of the performance of 540 development projects, they state that both the balanced matrix and the project matrix compared favourably with project teams in terms of cost, schedule and technical performance. All three structures shared roughly equal success rates (Figure 7.2). In contrast to studies from earlier decades (cf. Marquis and Straight, 1965), they found that these structures also outperform the functionally orientated structures in technical performance. If managers had do it all over again, almost half of them (45 per cent of the managers of all 540 projects) claimed that they would use a project matrix structure. This structure received the strongest recommendation from managers and was even better regarded than the project team approach (only 22 per cent of the managers recommended a project team). Hardly any managers (only 3 per cent) would use the functional approach if given a second chance. Project teams appeared to be better suited to very complex projects. This leads to the following hypothesis:

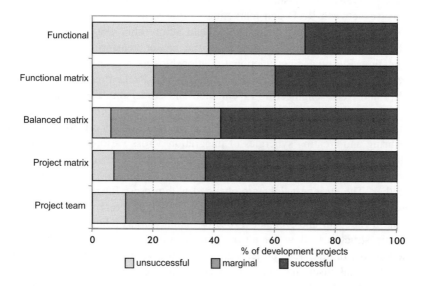

Source: Larson and Gobeli (1988, p. 186)

Figure 7.2 Overall performance

Hypothesis 7.1

Companies using a project-orientated approach towards innovation are more likely to be successful innovators than companies using functionally dominated structures.

It is important in this connection for the organisation to be capable of structuring itself laterally. Knowledge should not only be acquired in depth, but should also serve as a source of inspiration 'elsewhere' (De Bono, 1971), whether in other organisational departments or outside the organisation. New ideas are generated by exploring entirely new ground. The manner in which the company is organised strongly determines the possibilities for lateral relationships and the ease with which such relationships can be effected. Whereas the previous sections (and hypothesis 7.1) were focused on lateral organisation at company level, the next subsection focuses on lateral organisation at project level.

7.1.5 Multidisciplinary team approaches

Research has shown the importance of good cross-functional interfaces. Souder (1988), for instance, has shown that maintaining an effective interface between R&D and marketing can be vital for the success of new products. And according to Gerstenfeld and Sumiyoshi (1980), an inadequate R&D–marketing interface is one of the major reasons for innovative stagnation. But other interfunctional relationships have been studied as well: R&D–manufacturing (Clare and Sanford, 1984); R&D–finance (Anderson, 1981) and R&D–engineering (Weinrauch and Anderson, 1982). According to Gehani (1992), a concurrent product development process requires a higher degree of integration across different parts of an organisation. By nurturing and creating synergies of integration across various organisational subunits, the process of new product development can be accelerated. Such horizontal links are reinforced by the manner in which project groups are put together. The strongest links are obtained within cross-functional (or multi-disciplinary) development teams (Whitney, 1988), also referred to as 'rugby teams' (Takeuchi and Nonaka, 1986) or 'skunkworks' (Quinn, 1985). Such organic structures allow for an integrated, holistic approach to innovation phenomena. In this view, projects are run by teams in which most (or all) of the functional disciplines needed to complete the project are represented. Leadership within the team can (but need not necessarily) shift according to the phase the project is going through (marketing, product and process design, plant design, production and distribution). Thus, for example, product designers stay aboard when the product is entering the factory to learn how their design works out in the reality of production. Process engineers are involved when the first sketches by the product designers appear on the drawing board to anticipate the consequences for process design. In this way, the most important condition for learning is met, that is, the closing of the feedback loop (Sitter, 1994). The synergy within these teams arises from the linking of various disciplines, which enables effective feedback at an early stage. Since a large part of the control capacities are delegated to the group, interface problems will be reduced to a minimum. In addition, such groups provide more flexibility and capacity to improvise, which firms need in order to face the inherent insecurities and complexities of innovation processes.

The multidisciplinary team approach allows people from various disciplines to learn from each other's experiences, tune in to each other and find out more about the downstream consequences of upstream choices. Marketeers and production managers can confront designers and researchers at an early phase with problems which might arise in later phases as a result of the decisions they make. In time, this may very well lead to a reduction of feedback to previous phases. The importance of cross-functional co-operation for innovative success has been shown repeatedly. Many studies on multidisciplinary team approaches (for example, Mabert et al., 1992; McDonough and Barczak, 1991; Perry, 1990) indicate that empowered cross-functional teams with strong, dedicated leaders are important keys to success and shorter lead times. Pinto and Pinto's (1990) study on multilateral relations in 72 projects demonstrated that teams which exhibit higher levels of cross-functional co-operation have a significantly higher incidence of project success than teams with low cross-functional co-operation. Important benefits resulting from the use of cross-functional teams include (Henke et al., 1993): reducing product development time; improving product development outcomes; and reducing the hierarchical information overload at higher levels. Furthermore, cutting across traditional vertical lines of authority and responsibility allows companies to overcome some of the shortcomings of hierarchical structures. And the chance of higher-quality decisions will undoubtedly improve as a result of multifunctional inputs to the decision-making process.

Based on an extensive study of 103 new product projects in the chemical industry, Cooper and Kleinschmidt (1994) conclude that the strongest drive to ensure well-timed projects is provided by the use of a cross-functional, dedicated, accountable team, with a strong leader and top management support. It is therefore to be expected that front-runners will make more use of multidisciplinary teams than will pack members. Furthermore, it is to be expected that the use of multidisciplinary teams will lead to shorter lead times for innovation projects. Therefore, it was tested whether a multidisciplinary, project-wise approach would lead to increased speed in completing innovation projects:

Hypothesis 7.2
Companies which use multidisciplinary teams innovate faster (7.2a) and are more successful in their innovative efforts (7.2b) than companies which employ a linear approach.

7.1.6 Organisational culture

In their attempts to understand why a company's working organisation does not naturally adapt to changing circumstances, Burns and Stalker (1961, p. 138) found, somewhat to their surprise, that 'irreconcilable differences in attitude and codes of rational conduct' appeared to be important predictors in this respect. These irreconcilable differences can be regarded as concerning the culture of the organisation and the organisational subcultures. Organisational culture is therefore a factor to take into account in explaining innovative success. Before discussing recent literature on the relationship between culture and innovative success, we shall first define culture in general.

'Culture' can be regarded as the composite of a number of variables, including the type of leadership, prevailing stories and myths, accepted rituals and symbols, the type of power structure, the form of organisational structure, the decision-making process, functional policies and management systems (Johnson and Scholes, 1993). Organisational culture can therefore be defined as 'the deeper level of basic assumptions and beliefs shared by members of the organisation which operate unconsciously and define an organisation's view of itself and its environment in a "taken-for-granted" fashion' (Schein, 1989) and which will 'be taught to new members as the correct way to perceive, think, and feel in relation to the problems of survival and integration' (Schein, 1988, p. 8). Culture manifests itself in overt behaviours, norms and espoused values and can be regarded as the unseen and unobservable force that is always behind an organisation's tangible activities (Schein, 1985). Or as Kilmann et al. (1985) put it: 'Culture is to the organisation what personality is to the individual – a hidden yet unifying theme that provides meaning, direction, and mobilization.'

The influence of culture on management can be shown through its role in affecting the manager's perception of and response to changes in the internal and external environments (Sweeney and Hardaker, 1994). Culture affects the ways in which people consciously and subconsciously think, make decisions, and ultimately the ways in which they perceive, feel and act towards opportunities and threats presented by the internal and external environments. Deal and Kennedy (1982) originated the term 'strong culture' to describe companies which have widely shared values among organisation members. In these companies, employees from top to bottom share a collective vision of 'what the business is all about', as well as deep personal convictions that 'the company's success is my business' (Jelinek and Bird Schoonhoven, 1990, p. 368). Some studies claim direct linkages between strong culture and performance. Based on a study of 80 companies, Deal and Kennedy (1982) conclude that companies with strong cultures were also 'uniformly outstanding performers'. Peters and Waterman (1982, p. 75) conclude that 'without exception, the dominance and coherence of culture proved to be an essential quality of the excellent companies'. But the advantages of strong cultures are not unalloyed, as research by Jelinek and Bird Schoonhoven (1990, p. 365) shows.[4] In general, they indicate three broad categories of costs associated with strong cultures: individual (such as risk of burnout); managerial; and organisational (for example, a strong culture may blind the company to dramatic changes in the environment). Their message is that strong cultures have to be managed:

> People in the innovative companies we researched were keenly aware of the costs as well as the benefits. Employees in our sample companies praise their companies' cultures, value them deeply, and see many advantages to them. But people within these cultures also bear their costs, and invest much effort to avoid the potential hazards of a strong culture (Jelinek and Bird Schoonhoven, 1990, p. 369)

According to Schein (1988), cultures in which the invention of environmentally responsive solutions (as opposed to self-preservation) functions as the learning dynamic are among the organisational cultures which favour innovativeness. To identify factors explaining successful management of processes of change, then, it is important to gain an insight into the culture of the organisation (Hofstede et al., 1993; Sweeney and Hardaker, 1994). Organisational culture is therefore a factor to be taken into account in explaining innovative success. Ackroyd and Crowdy (1990), for instance, argue that the nature and intensity of an organisation's culture appears to be an important predictor in explaining differences in competitive advantage. Regarding the relation between organisational culture and innovative success, Moss Kanter (1983) differentiates between segmentalist management cultures and integrative cultures. She argues:

> The highest proportion of entrepreneurial accomplishments is found in the companies that are least segmented and segmentalist, companies that instead have integrative structures and cultures emphasizing pride, commitment, collaboration, and teamwork. (Moss Kanter, 1984, p. 178)

Jelinek and Bird Schoonhoven (1990, p. 381) identify clear links between informality or freedom and innovation. Informality facilitates a form of interaction among organisation members that deliberately and explicitly de-emphasizes organisational status in favour of task-relevant expertise. When de-emphasised status differentials and equality are reinforced by norms of informal communication, managers in the innovative companies they studied are fundamentally and far more explicitly dependent on their subordinates than is generally acknowledged in more traditional firms. Informal cultures also tend to stimulate creativity: 'An important outcome of informality and open communication is the constant expectation that those with ideas will speak up' (ibid. p. 384). It is through informal interchanges among peers, across departments, top-down and bottom-up, that new ideas can emerge. Furthermore, it has been shown (Ven, 1986) that some successfully innovating companies have developed tolerance and protective policies for deviant, non-conforming individuals whose social space provides a latitude for creativity, and that a culture based on innovation and entrepreneurship is stimulating to innovation.

In the light of these discussions, it is to be expected that certain organisational cultures favour innovation and innovative success more than others. We therefore expect to find sharp contrasts between front-runners and pack members regarding their company cultures and hypothesise as follows:

Hypothesis 7.3
The capacity of a company to innovate successfully will increase to the extent that its culture can be characterised as stimulating creativity, non-bureaucratic and progressive.

'Stimulating creativity' can be measured by the extent to which members of the organisation perceive their company culture as accepting of uncertainty, open to change and tolerant of failure. 'Non-bureaucratic' can be measured by the extent to which the company culture can be regarded as improvising, informal, status-inert and with few hierarchical layers. 'Progressive' can be measured by the extent to which the organisation is regarded as non-conformist and non-conservative.

7.2   RESULTS

7.2.1   Project structures

Hypothesis 7.1 argues that companies using a project-orientated approach towards innovation are more likely to be successful innovators than companies using functionally dominated structures. Respondents were therefore asked to describe the ways in which innovation projects were normally organised in their companies and to indicate which of the five basic structures best describes their approach. As Table 7.2 shows, front-runners tend to use a more project-orientated approach towards innovation management whereas pack members are more functionally orientated. Almost three-fifths (59 per cent) of the front-runners use a project-dominated structure compared with only 17 per cent of the pack members. From the latter group, more than two-thirds of the companies use a functionally orientated approach. The differences are even clearer when looking at the individual twins. We then see that in 68 per cent of the cases the front-runner twin uses a more project-orientated approach than its *respective* partner from the pack.

Table 7.2 Organisation of innovation processes (%)

	Pack members	Front-runners
Functional organisation	41	3
Functional matrix	28	24
Balanced matrix	14	15
Project matrix	17	41
Project team	0	18

The balanced matrix form does not seem to be used frequently. Only 15 per cent of all companies claim to use this as the most dominant form. This underlines the criticism of the balanced matrix discussed in this section. The single most widely used structure among front-runners is the project matrix approach. These findings underline the results of Larson and Gobeli (1988), who found that the project matrix received a stronger recommendation from managers than the project team.

As discussed in this section, the five basic structures can be regarded as discrete points on a continuous scale indicating the shifting balance of power between the functional manager and the project manager (and thus ranging from line organisation to full project organisation). The correlation between this scale and innovative success proved to be very strong (the correlation coefficient is 0.54) and statistically very significant ($p < 0.001$).

These results thus provide strong support for Hypothesis 7.1, which asserted that companies using a project-orientated approach towards innovation are more likely to be successful innovators than companies using functionally dominated structures.

Contextual data

During the start-up phase of our research project, there was a growing awareness in the literature that organisations experimenting with multidisciplinary teams encountered difficulties in reaping the full benefits of this approach (Clark and Wheelright, 1993, p. 522). Simply designating the team as 'cross-functional' or 'multidisciplinary' and setting up a regular schedule of meetings was not enough. In fact, as Clark and Wheelright state: 'many companies which opted for the team structure found that their managers were increasingly stretched, with less time for substantive work, and that projects took just as long –if not longer– to complete'.

Reasons for this lack of success can be attributed to the way the project was embedded in the organisation and the prevalence of 'old' cultures, but also to the weight of the project manager. Clark and Wheelright argue that, in contrast to project managers operating in functionally orientated structures and the balanced matrix, the project manager in the project matrix and project team structures should be a 'heavyweight' who exerts direct, integrating influence across all functions. They have come across such managers in the successfully innovating companies they studied and consider them to be heavyweights in two respects (Clark and Wheelright, 1993, p. 527). In the first place, they are senior managers in the organisation, sometimes even outranking the functional managers. They therefore have not only experience and expertise, but also hierarchical authority. Second, heavyweight project managers have primary influence over the people working on the innovation project and they supervise their work directly or through key functional members of the core team. In the project team structure, the core group is sometimes even physically located with the project manager. The project managers in the functional matrix and balanced matrix structures generally have less authority and are less senior than the heavyweight project managers. The main job of such 'lightweight' project managers is to co-ordinate the activities of the different functions. They generally do not have the power to reassign personnel or reallocate resources (Clark and Wheelright, 1993, p. 526).

In order to gain more insight into the extent of the project manager's influence on the project, the respondents were asked to indicate which responsibilities (from a precoded list of eight) are generally held by the project manager or project co-ordinator. Figure 7.3 shows that in more than half of the companies, the project manager is responsible for cost control, capacity allocation and technical design. Project managers in pack members bear more responsibility for personnel policy than project managers in front-runners. A possible explanation for this is the fact that they generally work in a more functionally orientated organisation in which project management and group management are often combined. This is also apparent from the fact that more than three-quarters of the pack members indicating that their project managers are responsible for personnel policy have an organisational form in which the functional managers have distinct influence (functional organisation, functional matrix and balanced matrix). In contrast, only one-third of the front-runners which make their project managers responsible for personnel policy have a functionally dominated structure.

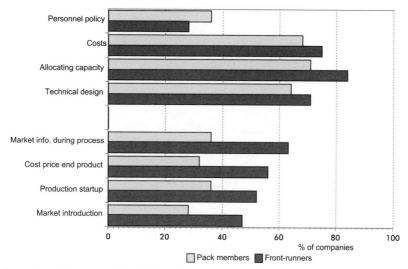

Figure 7.3 Responsibilities of the project manager

Furthermore, Figure 7.3 and Table 7.3 show that front-runners do not differ much from pack members with respect to responsibilities early in the project's life, such as personnel policy, project cost estimates, capacity allocation and technical design. These differences do not prove to be statistically significant (Table 7.3). However, front-runners do differ significantly from pack members regarding responsibilities in later phases of the project ($t = -2.93$, $p = 0.005$). These are responsibilities concerning activities undertaken when the project is nearing completion, such as the provision of market information during the project, establishing the cost price of the final product, production start-up and market introduction. From this we might conclude that project managers of front-runners tend to have more influence than their counterparts in pack members and that this difference is most specifically related to the later stages of the innovation process. In general, front-runner project managers are also responsible for more tasks than their pack-member counterparts, as Table 7.3 shows. The average front-runner project manager has 4.4 out of the eight listed responsibilities (against 3.2 responsibilities for the average pack-member project manager). It could be said that the project manager of a front-runner seems to be more of a heavyweight than the pack-member counterpart.

Table 7.3 Number of responsibilities of project managers

Scale (and variables from which the scale is derived)	Average pack members	Average front-runners	T-value
Responsibilities early in the project (personnel policy, costs, capacity allocation, technical design)	2.0	2.4	–1.04
Responsibilities late in the project (market information during the process, cost price of final product, production start-up, market introduction)	1.1	2.0	–2.93***
All eight responsibilities	3.2	4.4	–2.11*

Notes: * $\alpha \leq 0.05$; ** $\alpha \leq 0.01$; *** $\alpha \leq 0.005$ (one-tailed)

7.2.2   Cross-functional co-operation

Multidisciplinary development teams make it possible to achieve good co-ordination between departments. The question, however (Hypothesis 7.2a), is whether better co-ordination also leads to a faster innovation process. This question will be dealt with in subsection 7.2.5. We first examine the differences between front-runners and pack members in the use of multidisciplinary teams for innovative success (Hypothesis 7.2b). In addition we also explore the importance of the various bilateral relationships for innovative success.

It was clear from the previous subsection that front-runner organisations primarily follow the horizontal flow of ideas which ultimately result in new products, processes or services. Pack members tend to keep their innovative efforts within the traditional, vertical, functional limits. This trend is also seen in the frequency with which companies make use of multidisciplinary teams. Front-runners appear to make considerably more use of multidisciplinary teams for innovations than pack members ($t = -3.02$, $p < 0.001$), thus indicating a positive correlation between the use of multi-disciplinary teams and innovative success (as is argued by Hypothesis 7.2b).

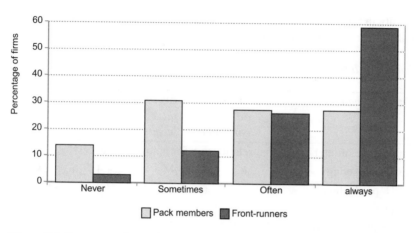

Figure 7.4 Frequency of use of multidisciplinary development teams

Figure 7.4 shows that 85 per cent of the front-runners often or always make use of multidisciplinary teams while innovating, as opposed to only 45 per cent of the pack members. The particular importance of multidisciplinary teams is that they bring people from different backgrounds (that is, from different, often functional, departments) into direct daily contact with each other and enable them to learn from each other.

Although various disciplines are involved in innovation processes, not all relationships are of equal importance to the success of such processes. Some relationships are so crucial for success that the chance of success with a poor relationship is slight. There are also relationships which, while definitely important, are not crucial. If such relationships are poor they might, for example, lead to considerable delays. A gradation can thus be indicated in the importance of the interfunctional relationships with respect to a given process. For each company, we asked all four respondents to indicate how important they consider a number of relationships to be for innovative success in their companies (regardless of whether or not these relationships are represented in the teams). These answers were then averaged per company. Table 7.4 shows the results of a T-test.

The most important relationships appear to involve R&D, with R&D–marketing as the most important, followed at some distance by R&D–production and R&D–engineering. Relationships involving the maintenance department, in contrast, were considered relatively unimportant to innovative success. Front-runners differ hardly at all from pack members in this respect. Although front-runners give more importance to the R&D–marketing relationship than pack members do, both groups score this relationship as the most important.[5] Moreover, almost half the front-runners state that high quality in this relationship is crucial for innovative success (= score 5), while only 28 per cent of the pack members share this opinion.

Table 7.4 Inter-functional relationships

Variable/scale	Average all companies	Average pack members	Average Front-runners	T-value
Importance of inter-functional relationship for innovative success[¶]				
R&D–marketing	4.2	4.0	4.4	–2.06*
R&D–production	3.8	3.7	3.8	–0.63
R&D–engineering	3.7	3.6	3.9	–1.37*
Engineering – production	3.7	3.7	3.8	–0.44
Production–marketing	3.3	3.5	3.1	1.68*
Production–maintenance	2.8	2.9	2.7	0.79
Engineering–maintenance	2.5	2.7	2.4	1.00
R&D–maintenance	2.2	2.2	2.2	0.03
Need for improvement in interfunctional relationships[†]				
R&D–marketing	2.9	2.9	2.8	0.10
R&D–production	2.8	3.0	2.7	1.54
Production–marketing	2.7	2.7	2.6	0.41
R&D–engineering	2.6	2.6	2.6	–0.05
Engineering–production	2.6	2.7	2.5	1.32
Engineering–maintenance	2.2	2.2	2.2	0.27
Production–maintenance	2.2	2.4	2.1	1.75*
R&D–maintenance	2.0	1.9	2.0	–0.36

Notes: * ($\leq$ 0.05; ** ($\leq$ 0.01; *** ($\leq$ 0.005 (one-tailed); ¶: 1 = unimportant; 2 = somewhat important; 3 = important; 4 = very important; 5 = absolutely crucial; †: 1 = no improvement needed; 2 = hardly any improvement needed; 3 = improvement needed; 4 = much; 5 = very much

If we also examine the extent to which the respondents feel that the inter-functional relationships in their companies need improvement, it appears that the most important relationships for success are also the ones which need the most improvement, while the less important relationships need hardly any improvement at all (Table 7.4).

Now that we have shown differences between front-runners and pack members regarding the use of multidisciplinary teams and the importance they attribute to the various functional relationships, we will test hypothesis 7.2a, which postulates that companies using multidisciplinary teams innovate faster than companies which do not. In total, 47 companies provided us with data about lead times of innovation processes. The average length of innovation processes at these companies was 16 months, with that of pack members being even slightly lower (14.5 months) than that of front-runners (17.6 months), although these differences were not statistically significant.

As Table 7.5 shows, we have found no conclusive evidence for this hypothesis in our data.[6] We did find, however, a correlation between the average length of the innovation processes and the extent to which the company makes use of multidisciplinary teams. But, contrary to expectations, this correlation proves to be positive. Furthermore, we did not find a negative correlation (as would be expected) between the average lead time of innovation projects and the extent to which innovation processes are organised project-wise. Instead, the correlation is positive, yet not significant. In the discussion section of this chapter we shall come back to these findings (subsection 7.3.2).

Table 7.5   Correlations with the speed of innovation (only the 24 full pairs were used in this analysis)

Independent variable	1	2	3	4	5
1. Average length of innovation processes	1.00				
2. Project-wise approach to innovation	0.17	1.00			
3. Use of multidisciplinary teams	0.28*	0.59***	1.00		
4. Success	0.13	0.56***	0.47***	1.00	
5. Innovativeness of company	–0.02	0.19	0.26*	0.52***	1.00

Notes: * $\alpha \leq 0.05$; ** $\alpha \leq 0.01$; *** $\alpha \leq 0.005$ (one-tailed)

7.2.3 Organisational culture

All respondents (approximately 240) scored their company's organisational culture on a 12-point semantic differential scale. For each company, the answers of the four respondents were averaged to achieve a more balanced representation of the firm's culture. The image of the average company that emerges is that of a result-orientated, flat and fairly informal organisation with relatively few differences in status. The average company is open to change, moderately progressive and tolerates failures, provided people learn from them.

In Table 7.6, four scales were calculated on the basis of the individual variables. The 'stimulating creativity' scale (comprised of variables indicating a culture characterised by low uncertainty avoidance, openness to change and toleration of failures) appears to correlate significantly with innovative success. So does the 'progressive' scale (non-conformist and non-conservative). The 'non-bureaucratic' scale showed no correlation with innovative success and Cronbach's alpha of the 'co-operation' scale was too low for reliability.

Table 7.6 Correlations between cultural dimensions and innovative success

Scale	Scales or variables used to construct the scale	Cronbach's alpha[a]	Correlation with innovative success
Stimulating creativity scale		0.5	0.24[a]
	Accepting uncertainty		0.20
	Open to change		0.30**
	Tolerant of failure		–0.06
Non-bureaucratic scale		0.7	–0.09
	Improvising		–0.05
	Informal		–0.21**
	Flat organisation		0.04
	Status-inert		–0.06
Progressive scale		0.6	0.26*
	Non-conformist		0.16
	Innovating/non-conservative		0.28*
Co-operation scale		0.3	n.c.
	Collectivist		–0.04
	Co-operative		0.01

Notes: * $\alpha \leq 0.05$; ** $\alpha \leq 0.01$; *** ≤ 0.001 (one-tailed); n.c. = not calculated because Cronbach's alpha < 0.5. [a]The 'stimulating creativity' scale should be interpreted with caution as Cronbach's alpha is only 0.5.

An examination of the individual cultural variables revealed that only three of them proved to be significantly correlated with innovative success. Front-runners appear to be more open to change, more formal and less conservative than pack members (Table 7.6). However, the differences between the two groups only revealed themselves in the magnitude of the characteristics, not in the opposing characteristics of the semantic differential (for example, structured as opposed to improvising). There was no semantic differential on which the front-runners generally scored opposite to the pack members.

We can thus state that only partial support was found for Hypothesis 7.3, which postulated that the capacity of a company to innovate successfully increases to the extent that its culture is characterised as stimulating to creativity, non-bureaucratic and progressive. Some evidence has been found to support the assumptions about creative atmosphere and progressive culture, but no evidence was found to support the proposition that successfully innovating companies are less bureaucratic than less successfully innovating companies.

As the lack of differences between front-runners and pack members surprised us, we examined whether these differences would become larger once the 'mid-field players' (the top pack members and bottom front-runners) were omitted from the analysis. This proved to be the case. When the whole sample was divided into three groups according to increasing innovative success and only the top 20 and bottom 20 companies were included in the analysis, the differences between top and bottom did become larger, but only the same three variables as before proved to be significant (Figure 7.5). A comparison of the top 12 against the bottom 12 showed even larger differences, yet still only the 'open to change', 'non-conservative/innovative' and 'accepting uncertainty' variables remained significant at the level $\alpha < 0.01$.

7.2.4 Contextual data

Setting goals and planning

Planning is often regarded as the key to successful project completion. Work cannot be controlled or managed unless it has been adequately planned. Delays and budget and schedule overruns occur for a variety of reasons, some of which can be addressed by proper planning. But to what extent can one plan for innovation? To what extent is the innovation process so stochastic and uncertain that one can only provide conditions instead of planning? Based on a study among 360 new product managers in 52 high-technology companies, Thamhain (1990, p. 15) found indications that a clear goal and effective planning correlated positively with team effectiveness: 'Comprehensive project development planning early in the project life-cycle has a favorable impact in the work environment and team effectiveness.'

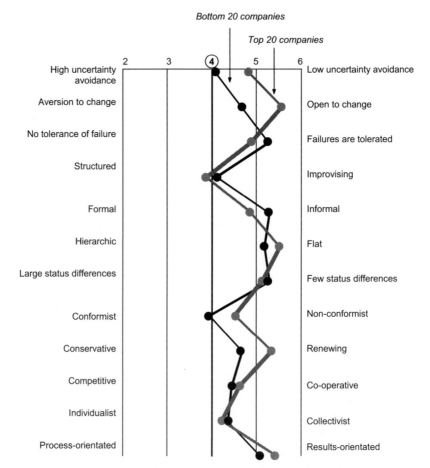

Figure 7.5   (Lack of) cultural differences between front-runners and pack members

Proper planning requires the participation of the entire team, including support departments, internal information suppliers, subcontractors and management. All these actors must show their commitment to the project and the planning. This then makes it easier for project managers to integrate the various tasks across functional lines in later phases. The question, then, is whether planning properly and setting goals at the start of the project is likely to lead to success. We used our data to see whether this holds true for a non-sector-specific sample as well. Our intention was to find out whether front-runner innovation projects are better planned at the start of the project and provided with clearer targets or goals than pack-member innovation projects. The respondents were thus asked how specifically the company plans its innovation projects and how specifically goals are set at the beginning of the project.

We found that, on average, technical and marketing goals are the two most clearly described targets in the initial planning document (Table 7.7). On the whole, front-runners and pack members do not differ in the extent to which these two goals are included in the planning. Front-runners, however, did differ significantly from pack members (see Table 7.7) regarding the specification of the profit goals ($t = -2.65$, $p < 0.005$), turnover goals ($t = -2.08$, $p < 0.05$) and cost objectives ($t = -1.38$, $p < 0.10$). Front-runners provide more direction to the teams regarding the cash flows generated by innovation than is the case with pack members.

Table 7.7 also shows that the time schedule is usually described concretely by front-runners and pack members. Also, the people who are to be involved in the project are usually clearly designated at the beginning of the project. While we saw earlier that front-runners usually described project goals better than pack members, it now appears that pack members devote more attention to the concrete description of the project planning itself. These differences, however, are not statistically significant.

Table 7.7 T-tests with respect to the specificity of planning

	Average pack member	Average front-runners	T-value
Setting goals†			
Technological goals	4.03	4.12	−0.30
Marketing goals	3.74	4.06	−1.19
Cost objectives	3.59	3.99	−1.38
Turnover goals	3.00	3.68	−2.08*
Profit goals	2.59	3.47	−2.65***
Planning†			
Time schedule	4.41	4.27	0.70
Persons involved	4.16	4.03	0.48
Bail-out points	3.76	3.66	0.28
Priorities to other projects	3.74	3.35	1.25

Notes: * $\alpha \leq 0.05$; ** $\alpha \leq 0.01$; *** $\alpha \leq 0.001$ (one-tailed); †: 1 = not mentioned in initial planning document ; 3 = broadly described; 5 = described with specific targets

We have also analysed the extent to which more highly specified turnover goals, profit goals and cost targets also lead to more stringent project monitoring and evaluation. As Table 7.8 shows, the more cost targets were specified, the more important their role became in steering and monitoring projects and evaluating the performance of the project manager. Furthermore, there was no correlation between the initial specificity of turnover and profit goals and the evaluation of the project managers' performance. A cautious conclusion might then be: if concrete aspects are stipulated in the project planning, it will be the cost targets in particular on which project managers are evaluated once the project has been completed. Meeting cost targets was therefore very important for their track records.

Table 7.8   Correlation between goal specificity in the initial planning document and the use of that document during the innovation process

	Project steering	*Interim progress evaluation and monitoring*	*Evaluation of project manager's performance*
Turnover goals	0.20	0.31*	0.02
Profit goals	0.25*	0.33*	
Cost targets	0.38**	0.37**	0.24*

Notes: * $\alpha \leq 0.05$; ** $\alpha \leq 0.01$ (one-tailed)

Planning also plays a considerably more important role in the later stages of the innovation process in companies which employ a project-wise approach than in companies which follow a more functional approach ($r = 0.40$, $p < 0.01$). Planning in general is also more concretely described in project-orientated companies than in functionally orientated companies ($r = 0.34$, $p < 0.05$). This is of course inherent in the project management discipline.

Cross-functional career patterns
From the previous analyses, it appears that front-runners differ from pack members primarily by organising more horizontally. This horizontal, stream-like innovation project focus can be given form in many ways; for example, in cross-functional career patterns. A transition from one function to another implies a transfer of knowledge and experience between functional groups. Such a career move can be made downstream (for example, from R&D to production to marketing) or upstream (from marketing to production to

R&D). It was tested whether the greater attention front-runners give to horizontal organisation processes could also be traced in career development. As can be seen in Table 7.9, cross-functional career patterns generally do not occur very often (2 = rarely), but they do occur more often with front-runners than with pack members. There are thus indications that successfully innovating companies pay more attention to cross-functional career patterns than do less successfully innovating companies.

Table 7.9 Cross-functional career patterns

	Average all companies	Average pack member	Average front-runner	T-value
Frequency of cross-functional career patterns[a]				
From R&D towards production	2.1	1.7	2.5	−3.06***
From production towards R&D	2.1	1.9	2.3	−1.57
From R&D towards marketing	2.0	1.7	2.3	−2.11*
From marketing towards R&D	1.5	1.3	1.7	−2.12*

Notes: * $\alpha \leq 0.05$; ** $\alpha \leq 0.01$; *** $\alpha \leq 0.005$ (one-tailed); [a]: 1 = never; 2 = rarely; 3 = sometimes; 4 = often; 5 = very often

Downstream career movements appear to occur slightly more frequently than upstream, but on average they are also quite rare. Upstream movements can mainly be found where marketeers have in-depth knowledge about the core technologies; for instance, in the chemical industry, where marketeers are sometimes recruited from chemical engineers.

7.2.5 Organisational competencies and innovative success

In order to find support for Hypothesis 3.1, which stated that organisational competencies are important predictors of innovative success, the significantly correlating variables were used to construct a regression equation. A step-wise multiple regression analysis was employed, which allowed us to limit the number of variables in the final equation to a minimum. But since some variables are highly correlated with each other, they might fail to meet the tolerance criterion in the regression analysis. It was therefore decided to correct for multicolinearity by slightly changing the tolerance criteria[7]. We must, however, bear in mind that this might lead to unstable parameter estimates. But since at this stage the object of the regression analysis is to detect organisational variables which will be useful in predicting innovative success rather than to determine *the strength* of each of these variables, the instability of the coefficients is of little concern.[8] In order to improve the solidity of the measurements, scales were constructed from variables where appropriate (Table 7.10).

Table 7.10 Scales constructed from individual variables

Scale	Scales or variables used to construct the scale	Cronbach's alpha	Correlation with innovative success
Innovation-fostering culture		0.7	0.28**
	Stimulating creativity scale		
	Progressive scale		
Cross-functional career patterns		0.8	0.40***
	Career patterns from R&D to production		
	Career patterns from production to R&D		
	Career patterns from R&D to marketing		
	Career patterns from marketing to R&D		
Multidisciplinary project-wise approach		0.7	0.55***
	Project-wise approach		
	Use of multidisciplinary teams		
Heavyweight project manager		0.7	0.26**
	Number of specifically mentioned project manager responsibilities (out of a possible eight)		
Setting goals		0.8	0.28**
	Specifying profit goals in project planning		
	Specifying turnover goals in project planning		
	Specifying cost targets in project planning		

Notes: * $\alpha \le 0.05$; ** $\alpha \le 0.01$; *** $\alpha \le 0.001$ (one-tailed)

Table 7.11 shows the effects of organisational competencies on innovative success. The organisational competencies, as operationalised by this model, appear to explain 37 per cent of the variability in innovative success. Other scales and variables did not contribute further to the explanation of the variability.

Table 7.11 Stepwise multiple regression analysis towards innovative success on organisational competencies

Independent variable	B	Beta	t
Multidisciplinary project-wise approach	0.121	0.435	3.92***
Innovation-fostering culture	0.043	0.167	1.50
Cross-functional career patterns	0.071	0.258	2.26*
(Constant)	0.611		22.15***

Notes: $R^2 = 0.37$, $F(3,54) = 10.55$***; * $\alpha \leq 0.05$; ** $\alpha \leq 0.01$; *** $\alpha \leq 0.001$

While all these scales correlate significantly with innovation success, in the regression analysis it is primarily the degree to which a company evolves towards a multidisciplinary project-based organisation that is significantly related to innovation success. Also highly significant are the occurrence of cross-functional career paths. Although the 'innovation-fostering culture' correlates significantly with innovative success, it becomes clear that its effect in the regression model is not significant. This is probably due to the high and significant correlations between this scale and 'cross-functional career patterns' as well as 'multidisciplinary project-wise approach'. Other models with more or different organisational variables or scales did not lead to substantially higher determination coefficients. It was thus decided to limit the operationalisation of the organisational competencies to: the extent to which innovation processes are organised in a project-wise, multidisciplinary approach; the extent to which cross-functional career patterns occur in the organisation; and the extent to which its culture can be regarded as stimulating creativity and progressive.

The results demonstrated a coefficient of determination (R^2) of 0.37, suggesting that, while organisational competencies are an important predictor of innovative success, they are by no means exclusive.

7.3   DISCUSSION

7.3.1   Organising horizontally

The increasing integration of products and services, concern for trans-unit core competencies, the importance of cross-fertilisation in the development of new products, increasing globalisation and the increasing complexity of processes, products and services all play their part in making it increasingly important to be able to think and act laterally. In this connection, Hertog and Huizinga (1997) argue in favour of organisational instruments and structures which promote trans-unit knowledge transfer both within and between organisations. This capacity to develop, transfer and utilise knowledge across organisational boundaries is also referred to as the company's 'lateral organisational capability' (Galbraith, 1994). The more organisational cross-links that are possible or present in the company, the greater this capability will be. Our research shows that organisational competencies are closely related to the concept of lateral organisational capabilities. Multidisciplinary teams (which are in fact lateral groups), the project-wise approach to innovation projects, cross-functional career patterns and so forth can all be considered as instruments of lateral organisation. Front-runners use these instruments considerably more often than pack members.

Burns and Stalker's (1961) criticism of mechanistic cultures and their endemic problems and behaviours is therefore still of interest, since many organisations today still exhibit these problems. As we saw in section 7.1, modern approaches to management are cognisant of this and invest great effort to secure project-based forms of (temporary) organisational groupings such as matrix structures within the basic bureaucratic framework. The main theme to be deduced from more recent work (for example, Peters and Waterman, 1982; Mintzberg, 1983; Moss Kanter, 1983, 1988, 1989a; Jelinek and Bird Schoonhoven, 1990; Andreasen et al., 1995) is the same drive for organic organisational structures as that propagated by Burns and Stalker. Many of these studies were focused on large or high-tech companies. It has become clear from the present study that the presumptions about functional deconcentration, multidisciplinary team approaches and relative autonomy as important predictors for innovative success in large, high-tech companies also hold true for small and medium-sized companies in a wide range of low- and medium-tech industries and service sectors. In fact, the most significantly differentiating organisational variables appear to be related to functional deconcentration and streamlining: a project-wise, multi-disciplinary approach to innovation management appears to be an important non-sector-specific factor for innovative success.

There is a wide variety of practices by means of which companies can promote accessibility and informal communication among team members (Box 7.1). Cross-links (for steering, consultation and decision making) between the functions may be effected at different levels:

- *Management level* Often institutionalised in the form of a management team consisting of the director and the heads of the diverse functional departments.
- *Group manager level* Often institutionalised in the form of a regular meeting between the group managers.
- *Staff level* Within the project team carrying out the work or between important specialists from the various groups involved in the development project.

BOX 7.1 PROMOTING CROSS-FUNCTIONAL COMMUNICATION

There are a variety of ways to promote and improve cross-functional communication. One approach instituted at several front-runners (for example, companies 7, 8, 32, 39, 54) is to put team members from various functional departments literally together in one room for the duration of the project. According to these companies, physical proximity is an effective way of enhancing informal communication and co-operation as well as stimulating cross-functional learning.

One front-runner (number 2, a chemical company) even initiated a game of musical chairs at management level. When the procurement manager retired, the vacancy was filled by a marketing sector head, whose place in turn was filled by a senior researcher, while a senior researcher and the technical marketing sector head switched places as well. The initial results are so promising that management is thinking about having another round of musical chairs in a few years.

In some supply-sector companies we visited, people with technical backgrounds were sometimes working in the marketing and sales departments. They claimed to have better insight into technological possibilities than marketeers who had not had technical training and, partly due to this, to be able to evaluate more effectively for the company, customer requirements in terms of their merits.

Translating customer needs into concrete R&D questions was usually easier for them than for non-technical marketeers.

Some companies (for example, companies 1, 2, 32) have a system of traineeships, either voluntary or obligatory, between a number of related departments. Trainees work or carry out an important assignment in a different department for a certain period of time. One company even went so far as to forbid trainees to visit their own departments at all during this period.

Such structures may be structural or *ad hoc*. Multidisciplinary teams, for example, are *ad hoc* structures. We saw that these teams are much more frequently created by front-runners than by pack members (see Figure 7.4). As far as structural cross-links are concerned, it can be stated generally that such links in pack members are particularly limited to the management level. In some cases, companies opted for a rather too select team even at that level. For example, the composition of the innovation projects steering group in pack member 52 can only be called dubious (to put it mildly). Production, purchasing and logistics form no part of this steering committee because, as the director puts it, they 'would only put the brakes on new development'. Co-ordination problems with these departments only come up once the projects have been launched, so the company is regularly confronted with delays and co-ordination problems. Front-runners, on the other hand, more often have structural cross-links at group manager and staff levels and are more open to healthy confrontation between functional representatives.

Breaking through compartmentalisation thus appears to be an important organisational task in connection with the renewal of products, services and processes. Most companies have histories characterised by vertical structuring into functions and work areas, a heavy hierarchy and elaborate rules and procedures. They move away from this 'machine bureaucracy' model and make a 90-degree turn: processes are not to be managed via vertical but via horizontal lines. One could speak of stream-based organisation along the innovation chain. The product and the market are taken as the starting point, not the function. Front-runners have made more progress in this respect than pack members. The results provide a consistent picture. Horizontal, stream- and project-wise lines are drawn more distinctly by front-runners, while functional, vertical lines are drawn more distinctly by pack members. The results indicate that the strong horizontal lines found in front-runners usually concern the entire development process, from idea development and design to production and market introduction. In this way, upstream personnel (developers and designers) learn to better assess the consequences of their design decisions for those who have to work with their decisions downstream (in production and with customers). That is an essential condition for a learning organisation and for reducing the duration of the innovation process. Front-runner project managers carry more weight than their pack-member counterparts and their involvement does not stop with design but continues until implementation.

For front-runners, working with multidisciplinary teams is the rule rather than the exception. They usually ensure that all links in the innovation chain (from raw materials to customer) are represented on the teams. The criterion for team membership is the concrete input (specialty and so on) of the people involved, not their formal positions (Box 7.2). In this respect, it is important to recognise the truly essential interfaces. In particular, these appear to be those related to the R&D department.

Two important and closely related communication processes between two departments are 'information exchange' and 'confrontation'. Knowledge of the process as a whole is spread over different departments. In principle, each department has a monopoly on a piece of this knowledge. Through information exchange, these pieces can be put together to provide accurate insight into cause-and-effect relationships. In this way, upstream departments are confronted with the consequences of their decisions downstream in the process. Conversely, departments further downstream acquire insight into the reasons for certain upstream decisions. Companies have three available options for stimulating this information exchange: co-operation at project level (interdisciplinary teams); cross-functional career patterns; and task rotation.

BOX 7.2 PROJECT-WISE WORKING CAN MEAN DIFFERENT THINGS

For each relatively sizeable customer order, front-runner 54 (450 employees, contracting sector) sets up a multidisciplinary team. The project manager is selected from the functional area most closely related to the nature of the project. Project teams are put together across hierarchical structures, ensuring that different echelons are usually represented in the teams. Not infrequently, the project managers are from lower echelons in the organisation than some of their team members. The team is then assigned a room which will become the workplace of all team members for the duration of the project. Each team has a support group consisting of the director, the project manager's functional superior, the head of the sales department and the head of the calculation department. If specific problems arise in the course of the project, an R&D staffer will be called in to solve them. He or she will visit the customer to place the problem in its overall context.

Multidisciplinary teams are also formed by a pack member in the clothing industry (number 27). The management team of this company states that it does not often intervene and limits itself to giving advice. It seldom or never overrules the team. On the other hand, the director explains, it is clear to everyone that if a project goes really wrong the responsible project leader's job will be on the line. The marketing manager describes the management style as 'antiquated' and stuck within the existing structures. The management gives ambiguous signals. On the one hand, it wants to take a more project-wise approach, but on the other hand, it is wary of delegating and does little to actually implement project-wise working methods. According to the contact person, project managers have just as much authority as they grant themselves. But, he adds, 'when all is said and done, no one has any decision-making authority in our company except the two owner-managers'. The reality is thus different than the director would have us believe. The company has a number of promising young people who hardly dare to take any risks and are thus reluctant to start new things.

'The task of a project manager in our company is comparable to that of a football coach. He has to put the right people in the right place and see to it that they do what they are supposed to do'. (front-runner manager).

Co-operation at project level

Such co-operation may be both bilateral and multidisciplinary. With front-runners, we particularly encountered the latter form of communication. The multidisciplinary teams discussed above are examples in this respect. What we saw with pack members, however, was that interfaces were frequently focused on conflict rather than on the reduction of uncertainty. Departments often have an 'us-against-them' mentality which can be a serious obstacle that prevents people from really listening to and learning from each other (Box 7.3).

Cross-functional career patterns

Horizontal career patterns represent a second way of transferring knowledge and information between departments. As the analyses show, cross-functional career patterns were found much more frequently with front-runners than with pack members. The front-runner in the insurance business has a policy aimed at stimulating job transfers between the IT department and other departments. This doesn't seem to be easy. Switching from 'automation' to other departments is more frequent than switching from an operational department to an IT department. None the less, the management regards such changes as very important career moves. The added value of such career moves lies particularly in the fact that they make it possible to more effectively gear process innovations (originating in the automation department) to operational requirements. After all, people who once worked in operational departments are now helping tend the cradle of process innovation.

BOX 7.3 ANECDOTES: CROSS-FUNCTIONAL (NON-) CO-OPERATION

'The production department is there in the first place to make things, not to decide what things they are going to make' (R&D manager, pack member).

'According to marketing, more attention will have to be paid to the user-friendliness of the equipment in the future. I don't really believe in that. So it doesn't have any priority in our innovations. If we can make the equipment more user-friendly we will, of course, but we are not particularly looking for ways to do so.' (R&D manager, pack member).

'Multidisciplinary co-operation goes considerably further than maintaining bilateral contacts with different departments.' (Director, front-runner).

Task rotation

In addition to lateral career moves, temporary task rotation can also be employed to promote communication between departments, for example, traineeship 'exchanges' between departments. Thus, front-runner 46 has a policy of temporarily rotating employees between departments. In particular, people from the automation department follow process innovations down the line in order to ensure successful implementation.

It should be pointed out that the data used here concern primarily a 'snapshot' of a given moment in time. Most companies (the front-runners sooner and faster than the pack members) are going through an evolution from a vertically to a horizontally orientated organisation. Based on the qualitative material, it is possible to identify a clear trend. Moreover, there is also a distinct phase difference between front-runners and pack members in this respect (Box 7.4).

BOX 7.4 A PHASE DIFFERENCE

Company 47, an insurer, is highly functionally organised. The company has a number of management groups which to a large extent cover the same clients. They are aware that it is time to change course. The organisation is to be reorganised into business units serving defined market segments. The firm's major competitor had already turned its organisation upside-down years ago.

Company 2 is a producer of feedstock chemicals. It had already accepted the idea that more product-orientated and multi-disciplinary co-operation was necessary more than five years ago, but the policy teams put together with this in mind, made up of senior producers, developers and marketeers, weren't really getting anywhere. The teams became preoccupied with details of execution, and top-down communication remained much too time-consuming. The real breakthrough came by giving the co-operation concrete form through the work. For example, by giving multidisciplinary teams responsibility for upgrading new products with technical problems.

Company 34, a technical systems builder, has been working with multidisciplinary teams for a few years now. The surrounding organisation (mother and sister companies) still has a highly functional character. The few good project managers run from pillar to post and are dragged from one project to the next. It is almost impossible to make use of their experience in follow-up projects because they will be calling the shots in an entirely different project by that time. Company 39 also develops complex technical systems. Its project organisation is highly developed,

---

something which proved to be very much needed for forging close links between the functional groups involved with projects. In fact, it became so highly developed that project management turned into a successful business. The company now takes over the management of huge external projects and its internal projects benefit from the healthy crop of project managers as well.

7.3.2 Pace of innovation processes

Based on our results we cannot conclude that front-runners are faster innovators than pack members. The two groups did not differ significantly with respect to the average lead time of their innovation processes. Furthermore, contrary to expectations, the data even showed a positive correlation between the use of multidisciplinary teams and the length of the average lead time of innovation.

Basically, we can come up with two possible explanations for these results:

1. the multidisciplinary team approach is less effective and takes longer than assumed; or
2. the multidisciplinary approach is especially used to master more innovative or complex (and hence more time-consuming) projects.

Based on overwhelming evidence from other studies and above all when we examine the nature of the innovation projects and products in the companies in our study, we can safely state that the first explanation is not very likely. The most obvious explanation is thus the second: apparently, the type of project differs. It can be plausibly assumed that front-runners or multidisciplinary teams take longer on projects because the projects are more complex and require lengthier time horizons. After all, there will be more problems to be overcome in such projects. This way of looking is also reinforced by a number of other observations from our study:

- *Companies using a multidisciplinary approach are more innovative* A possible explanation might also be found by taking the innovativeness of the company into account. We have, for instance, found a high and significant correlation between innovativeness and the extent to which the company uses a project-wise approach. Furthermore, a positive correlation was found between innovativeness and the frequency of multidisciplinary approaches.

- *Companies using a project-wise approach experience fewer co-ordination problems between functional departments* It also appears from our study that a project-wise approach leads to a reduction in co-ordination problems between departments. Thus, the results show that companies which work with a project-wise approach less frequently encounter problems resulting from techn(olog)ical difficulties that arise in connection with the conceptualisation of an idea (Table 7.12).
- Companies using a multidisciplinary project-wise approach are more on time in the market with product innovations than companies using a functionally dominated approach. Companies using a multidisciplinary approach feel that they are more often on time completing projects than companies which rarely or never adopt such an approach. While innovation projects in companies which work with a project-wise approach may indeed take longer than projects in companies which operate functionally, the functionally operating companies are frequently much later to enter the market than the companies which work project-wise.
- Finally, the qualitative interviews reveal that *front-runners are generally very satisfied with the speed of innovation* and feel that they have the innovation process well in hand.

Table 7.12 Correlations between problems encountered while innovating and a project-wise multidisciplinary approach

	Project-wise approach	Multidisciplinary approach
Degree to which the following problems are encountered while innovating		
Inaccurate market analysis	−0.09	−0.18
Technical feasibility of idea	−0.24*	−0.19
Upscaling in production	−0.20	−0.05
Higher costs than expected	−0.17	−0.01
Reaction of competitors	−0.11	−0.05
Reaction of customers	0.12	0.02
Not enough time	−0.22*	−0.22*

Notes: *$\alpha \leq 0.05$; ** $\alpha \leq 0.01$; *** $\alpha \leq 0.001$ (one-tailed)

'Pace of innovation' is thus a relative concept. A duration of three years for a very creative product innovation may be 'faster' than a duration of two months for an incremental innovation. A careful tentative conclusion might therefore be that, for the companies studied, the multidisciplinary team approach to innovation did not pay off in terms of increased speed so much as in terms of greater innovativeness. Apparently, the explanation can be found in the interaction between organisational structures and the strategic choice to tackle complex innovation projects. A company which decides in favour of more complex innovation projects will also have to choose the best-equipped organisational structure for the purpose, that is, the multidisciplinary team approach. On the other hand, it may also be that organisations which follow a multidisciplinary team approach are able to tackle more complex and innovative projects for exactly that reason. Thus, although the relationship between the organisational competency which consists of the ability to follow a multidisciplinary team approach and the tendency to choose more complex and innovative projects can be demonstrated with reasonable certainty, the cause-and-effect relations are not clear. Perhaps this is a case of interaction between two mutually reinforcing processes.

7.3.3 Culture

The analyses support the propositions stating that an essential condition for effective innovation is to break through functional, bureaucratic structures and mechanistic cultures. Guided by recent discussions regarding company culture (Hofstede et al., 1993) and innovative culture (Jelinek and Bird Schoonhoven, 1990) and statements from managers (Box 7.5), we thus expected sharp contrasts between front-runners and pack members regarding their company cultures.

As Table 7.6 demonstrated, however, these were not found. On the contrary, one cultural dimension even scored the opposite of expectations. In contrast to what Hypothesis 7.4 suggests, pack members, for example, tend to score their companies as more informal than front-runners. A possible explanation for this is that highly bureaucratic organisations can usually only function thanks to an informal network in which problems which cannot be addressed within the formal system can be solved (Dalton, 1959). You have to know the 'shortcuts'.

BOX 7.5 THE IMPORTANCE OF CULTURE

The importance of company culture to innovative success was emphasised by one of the most successfully innovating companies in the sample (company 54). In this company, potential new employees are screened explicitly to determine whether they will fit into the company culture. A good deal of weight is given to this criterion. People who don't fit are not hired. New employees who prove to be unfit are dismissed after their probationary periods. The culture is described as dynamic, flexible, customer-orientated and open to change.

Company 2 (front-runner) realised eight years ago that it could never become successful unless a cultural reversal took place. The company had a highly compartmentalised organisation. Departments had no knowledge about each other and there were regular co-ordination problems. An important element in the process of change was thus to break down the walls and change from a culture based on risk avoidance to a culture in which individuals dared to take responsibilities (and thus risks). It was a long and tedious process, but it paid off in the end.

Regarding the overall lack of clear differences between the company culture as perceived by front-runners and pack members respectively, we have four possible explanations:

1. One explanation might be that the respondents referred to the culture of their department or organisational group instead of the culture of the organisation as a whole. Since organisations are composed from various subgroups, each with its own group culture, the total organisation may have both an overall culture as well as a set of subcultures. Therefore, every individual in the organisation will simultaneously possess elements of all the cultures of which he or she is a member (Maanen and Barley, 1984). Although respondents were explicitly asked to describe the culture of the company, it is possible that they found it difficult to distinguish between the 'group' culture and the company culture. It was therefore tested whether the three most homogeneous respondent groups (research managers, marketing managers and CEOs) differed from each other regarding their opinion of the organisational culture. In this way, it was possible to compare the evaluation of company culture by R&D managers with the views of marketing managers. However, the differences proved to be quite small and insignificant.

2. Another explanation might be that 'culture', as it was measured, is strongly correlated with organisational factors built into the matched-pairs design, such as the size of the company. To a certain extent, this proved to be the case. Some cultural dimensions showed stronger

correlations with the size of the company as measured by the number of employees (hierarchical culture and large status differences) or annual turnover (formal organisation, low tolerance of failure) than with innovative success.

3. A third explanation might be found in the instrument by which we measured culture. We used semantic differentials, but the results indicate that culture is more concerned with paradoxes. It has to do with and/and questions, not or/or ones. For instance, a probing, questioning culture is a prerequisite for successful strategic innovation. The existing conceptual map has to be challenged. Conversely, commitment to a single shared vision can unite and focus the firm's efforts. Achieving both is a matter of coping with paradoxes.

4. And finally, we must observe that the validity construct (the question as to whether or not a concept has been correctly operationalised) is always a thorny issue in research into organisational culture. In particular, it can be questioned whether a phenomenon as complex and implicit as organisational culture can be measured with simple rating scales (Hertog and Sluijs, 1995). Here again, the great value of the multimethod approach is evident. For the fact that culture matters emerges with particular clarity from the qualitative material. The quantitative material has little of value to add in this respect.

NOTES

1. We are presenting here a summary of the 11 characteristics listed by Burns and Stalker (1961) for each of the two structures.
2. The next step in managing the innovation processes more effectively was to break down the large divisions and create business units at the lowest possible organisational level, in which the management of marketing, engineering, product development and production become integrated in units with a clear responsibility for the business as a whole. As our study, however, focused on the level of the SME or business unit, we shall not look into these structural issues. In this book, rather, 'structural issues' are seen as relations *within* the business unit or SME.
3. In small companies, we sometimes see functional managers also acting as project managers. In such cases, Figure 7.1 should be seen primarily as an illustration of the two roles managers can have. Their authority as project managers goes beyond the limits of their functional role.
4. Jelinek and Bird Schoonhoven (1990) argue that 'strong culture' is often mistakenly defined as 'good culture' (Peters and Waterman, 1982; Deal and Kennedy, 1982). There are many relatively poorly performing companies which can be said to have a strong culture as well, and it is too simple just to define these as 'dysfunctional cultures'.
5. The same reasoning holds for the relationship between production and marketing, where the difference between front-runners and pack members is also significant but the relationship itself has the same place in the ranking.

6. In total, only 13 complete pairs had no missing values with respect to this correlation matrix. If only these complete pairs were taken into account (to correct for possible industry-specific lead times), the Pearson's correlation coefficient of variables 3 and 1 remained almost equal, as shown in Table 7.5: 0.26. But due to the smaller sample size, this correlation coefficient was no longer significant.
7. The maximum probability-of-F to enter the equation was set to 0.3.
8. To make the variables more comparable, it was decided to compute the regression model with the normalised values (as was also done in Chapters 5 and 6).

8. The competencies triad

Usually when we think about advanced civilizations we have in mind one that is thousands or millions of years ahead of us. But what worries me is a civilization that is just fifteen minutes ahead of us. Its members would always be first in line at the movies, and they would never be late for an appointment. (Woody Allen; cf. Moss Kanter, 1983, p. 64)

8.1 TWO REMAINING QUESTIONS

The previous chapters have shown us the ways in which front-runners are differentiated from pack members on each of the three competencies. It has been shown that differences in innovative success can be explained by differences in technological competencies, marketing competencies or organisational competencies. Two questions still remain unanswered: How important is the past in explaining current differences in innovative success? and: To what extent is a combination of strengths in competencies a better predictor for innovative success than strength in a single competency? The first question relates to the importance of the differences in timing of company policy and the development of competencies in the (recent) past in explaining innovative success. The second relates to the importance of a combination of compentencies.

8.1.1 Differences in timing of company policy

In this book, the explanation for innovative success has primarily been sought in the present. The survey provides a snapshot of a given moment in time, a look at the competencies present at that moment. In Chapters 5 to 7 inclusive we were looking mainly for covariances between the variables at that time. In this chapter, among other things, we want to examine whether a temporal dimension can be added, enabling us to not merely look at competencies but also to get a sense of the process which produced those competencies. The questions in that case are, for instance, whether one company is more successful than others because it began to build up competencies earlier or whether it gave priority to certain policy issues sooner than other companies. A matched-pairs study in the Netherlands on the differential performance of firms in bear markets (Schreuder et al., 1991) has shown for instance the importance of timing in company policy for the success of firms. Although the successful firms in their sample do not seem to differ much from unsuccessful firms in terms of forecasting ability, the successful firms did appear to act more vigorously when anticipating declining demands. Successful bear-fighting firms take *more* measures and take them *earlier* than unsuccessful firms. The actions taken by successful firms can be classified as anticipatory (in the areas of management, production scale and market), while the actions taken by the unsuccessful firms can be said to be more reactive to the bear markets.

Arthur (1989) argues that history matters, and that firms are 'locked in'. Path dependencies constrain a company's choices in the sense that where it can go is largely a function of where it has come from – the path it has followed, its current position and the paths or technological opportunities ahead. An explanation from the past can thus be sought in the policy being conducted. Managers have to divide their attention over numerous issues. Time, money and talent are scarce resources and choices have to be made as to what goals they should be aimed at. When these choices constitute a coherent whole and are aimed at the long term, we can speak of a 'company policy'. The question then concerns the extent to which differences in policy choices made in the recent past or in the timing of these policy choices can explain differences in innovative success between firms today.

As discussed in subsection 1.2.1, research has shown that the performance criteria on which firms compete have changed over time in a cumulative fashion (Bolwijn and Kumpe, 1989). Starting with effectively controllable and verifiable efficiency aspects, each subsequent factor became less controllable. The more controllable competitive factors such as efficiency, customer orientation, quality and service appear to have lost their competitive value as more and more firms are able to control them. The factor which currently has the highest competitive value, innovativeness, is the least controllable of them all (ibid.). It is therefore an illusion to think that routinisation and techniques which applied to the previous criteria are still adequate. The need to innovate effectively and efficiently confronted the organisations with new requirements. The question, then, is whether successfully innovating companies have responded to these new requirements earlier than less successfully innovating ones.

The previously mentioned matched-pairs study by Schreuder et al. (1991) showed statistically significant differences in the timing of organisational and managerial measures between firms which performed successfully on a bear market and firms which performed unsuccessfully. Our proposition, therefore, is that front-runner business policy has differed from that of pack members in the past decade in that the front-runners have paid more attention to organisational policy items than the pack members. Hence, we hypothesize that front-runners differ from pack members in the extent to which they gave attention to organisational policy issues in the past decade. One may assume that there is a phase difference not only with respect to organisational competencies but regarding marketing and technological competencies as well. At first glance, there often seemed to be no differences between front-runners and pack members: certainly not where the application of modern insights into marketing and R&D management is concerned. Apparently, such insights have penetrated the top half of the Dutch business community (front-runners and pack members) to such an extent that front-runners can no longer be distinguished from pack members on this basis. As was shown in the previous three chapters, front-runners indeed distinguish themselves from pack members with respect to a number of aspects which at the time of the study had hardly been discussed at all in the literature. The question which then arises is whether a significant explanation for the difference between front-runners and pack members can be found in the past, in the differences in timing of company policy. That is, do front-runner companies acquire their competitive edge because they change course sooner than pack members? On this basis, we have formulated an additional hypothesis.

Hypothesis 8.1
Front-runners differ from pack members in the timing of company policy with respect to the development of their competencies.

This hypothesis originated after the research was completed and the individual competencies had been analysed. As the data were already gathered when the hypothesis was constructed, we were limited to those data already collected for testing the hypothesis. As will be shown in the results section, this was especially problematic for the technological competencies. In the follow-up research projects, more retrospective variables will be included to enable a better test of this hypothesis.

8.1.2 Interaction between competencies

Until now, we have treated the three types of competencies individually. However, if we take a holistic or integrated approach towards innovation as the starting point, the question arises as to whether innovative success might actually depend on a combination of competencies. In the literature, the importance of an integrated approach (combining organisational, marketing and technological facets) in explaining corporate success has been extensively documented. Moss Kanter (1983, p. 27) found that the entrepreneurial spirit which generates innovations is associated with an 'integrative' way of approaching problems: the willingness to move beyond received wisdom and to combine ideas from unrelated sources. In an integrative climate, problems are seen and treated as 'wholes', and as related to larger wholes (the context). Such organisations reduce rancorous conflict and isolation between organisational units; create mechanisms for the exchange of information and ideas across organisational boundaries; and ensure that multiple perspectives will be taken into account in decision making. It is in such environments that innovation flourishes, according to Moss Kanter. On the other hand, companies which have adopted the contrasting management style, referred to as 'segmentalism', find it difficult to innovate or handle change. The segmentalist management style (Moss Kanter, 1983, p. 28) is concerned with compartmentalising actions, events and problems and with keeping each piece isolated from the rest. Problems are seen as narrowly as possible, independently of their contexts and relationships to other problems.

Grant (1991) argues that, for most firms, the most important capabilities are likely to be those which arise from an integration of functional capabilities, as the McDonald's example in Box 8.1 illustrates. Other findings suggest that adequate communication and co-ordination between marketing and technical areas is a key success factor for innovation projects (Calantone et al., 1993). Many studies (Cooper and de Brentani, 1984; Link, 1987; Crawford, 1991) suggest the existence of at least two main areas of competence in innovative success: marketing skills, resources and abilities, and technological skills, resources and abilities. A study by Larson and Gobeli (1988) identifies organisational structures, marketing competencies and technological competencies as antecedents of innovative success. Additional support was found by Calantone et al. (1993), who show that flexibility in organisational structure is related to higher levels of skills and resources in both marketing and technical areas. Articles on 74 companies rated by *Business Month* as among the five best-managed firms during each of the 15 years in the period 1972–86 were content-analysed by Varadarajan and Ramanujam (1990) in an effort to isolate the key strategic and organisational factors associated with superior corporate performance. Their main conclusion was that, far from being the result of applying a particular formula, superior performance requires a diverse mix of competencies and values. Furthermore, it has been shown that firms with well-developed supplier and customer relationships have better access to external technology (Mansfield, 1989; Urban and Hippel, 1989); and firms with well-established collaborative relationships between R&D and marketing (organisational competencies) are more likely to incorporate customer needs in the early stages of a product development process (Souder, 1988). Practical examples from the international business literature also point in the direction of an integrated approach to competencies (Box 8.1).

The question as to whether the combined effect of more than one competency provides a better explanation for innovative success has so far remained unanswered in this book. This question will be dealt with in the current chapter. It is our assumption (see Hypothesis 4.2) that a combination of competencies explains innovative success better than any of the single-competency models. In this chapter, we shall test the extent to which a combination of competencies can better explain differences in innovative success than any of the single competencies. The question, then, is whether one company is more successful than the other because it is strong in a combination of competencies while the other has mastered only one competency or perhaps none.

The three multiple regression models which resulted from these analyses did show significant relations between each of the competencies and innovative success, but none of them proved to be very strong. With an R^2 of the multiple regression equation of 0.41, the technological competencies appeared to be the best in explaining the variance in innovative success, but are still not strong predictors. An alternative explanation for innovative success might therefore be that front-runners excel in more competencies than pack members and that the combination of strong positions in two or three competencies is what makes a firm successfully innovative. The idea that innovativeness requires appropriate organisation, the ability to link to the market and appropriate technical skills has been established in a wide range of studies (for example, Link, 1987; Larson and Gobeli, 1988; Cooper and de Crawford, 1991; Calantone et al., 1993). These studies all refer to the importance of an integrated approach combining organisational, marketing and technological facets as a necessary condition for corporate success. However, many of these studies are focused on managing innovation projects and determining factors that account for the success of innovation projects (for example, Freeman, 1982; Cooper, 1986).

What we would like to test is whether such an integral approach also holds true with respect to the influences of the three competencies on a firm's innovative success. We thus want to test whether the three competencies in combination are important predictors for innovative success and whether front-runners excel in more competencies than pack members. This brings us back to the basic research hypothesis underlying this study as defined in Chapter 4. Hence:

BOX 8.1 ILLUSTRATIONS FROM PRACTICE

The New York Life Insurance Co. identified three key competencies that have determined its continued success: the ease with which its customers and sales agents could do business; support for its primary distributors; and swift response to customer desires (Gammill, 1992). In our framework, we would classify these as marketing competencies. However, these marketing competencies could only yield increased performance for the New York Life Insurance Co. after the firm had acquired organisational competencies as well. Changes in organisational behaviour were necessary: New York Life needed to break down barriers, cross departmental and functional lines, and integrate departmental functions.

Another example is given by the ADK Research Corporation, which used market research to link its innovation and core competencies to customer needs (Roth and Amoroso, 1993): 'Although potential customers did not explicitly value the technology itself, links could be made between technological competencies and more valued service dimensions such as communications flow, meeting deadlines, and staff responsiveness'. In our framework, the latter could be categorised under the heading of marketing competencies. Hence a combination of technological and marketing competencies.

Brockhoff and Pearson (1992) argue that the degree of technical and marketing aggressiveness has an effect on a firm's performance. They show that technology-based organisations can benefit from adopting a more aggressive marketing strategy, especially by striving to become market leaders in niche areas. Again, a combination of technological and marketing competencies.

McDonald's possesses outstanding functional capabilities in product development, market research, human resource management, financial control, and operations management. Critical to McDonald's success, however, is the integration of these functional capabilities to create the concern's remarkable consistency of products and services in thousands of restaurants spread across most of the globe (Grant, 1991, p. 121).

Hypothesis 4.2

If competencies are jointly present, firms are more likely to innovate successfully.

So we expect that front-runners are more likely to have strengths in more than one competency than pack members.

Table 8.1 Factor-derived scales regarding company policy

	Previous five years			Five to ten years ago		
	Average pack member	Average front-runner	T-value	Average pack member	Average front-runner	T-value
Innovation focus	3.3	3.9	3.38***	2.8	3.1	−1.45
Process innovations, product innovations, vocational training, hiring highly skilled people (Cronbach's alpha = 0.7)						
Streamlining	3.7	3.8	−0.67	2.7	3.2	−2.13*
Cost reductions, quality assurance, reorganisations, concentrating on core activities (Cronbach's alpha = 0.7)						
Market expansion	4.0	3.9	0.40	3.5	4.0	−1.78*
Increasing sales, enlarging market share (Cronbach's alpha = 0.9)						
Customizing	3.8	3.8	−0.03	2.5	3.2	−2.12*
Implementing new information systems	3.8	4.0	−0.71	2.4	3.1	−2.71***

Notes: * $\alpha \leq 0.05$; ** $\alpha \leq 0.01$; *** $\alpha \leq 0.001$ (one-tailed); 1 = top priority; 2 = high priority; 3 = medium priority; 4 = low priority; 5 = no attention devoted to

8.2   RESULTS

8.2.1 Policy agenda priorities

In order to gain insight into (the changes in) company policy over the ten years preceding the year in which the interviews were conducted, the general managers completed a questionnaire in which they indicated the importance of each of a list of 18 policy priorities over the previous five years[1] and the five years before that.[2] A factor analysis on these policy items revealed four factors. Scales derived from variables loading high on three of these factors proved to be reliable (as measured by Cronbach's alpha). These scales were labelled 'innovation focus', 'streamlining' and 'market expansion'. From the fourth factor, with the variables 'customising' and 'implementing new information systems' loading high, no reliable scale could be constructed. We therefore choose to include these two variables separately in the analysis. Table 8.1 provides the results of the T-tests.

In Figure 8.1 the differences in policy priorities between front-runners and pack members are shown graphically.

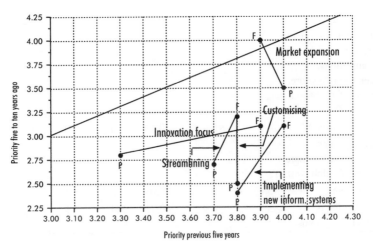

Figure 8.1 Differences in timing

On the horizontal axis, the priority of the policy items in the five years preceding the field research is depicted, while the vertical axis shows the priority attributed on that scale 5–10 years previously. The length of a line indicates the magnitude of difference between front-runners (F) and pack members (P), as Figure 8.2 explains.

The dotted lines (not shown in Figure 8.1) indicate the difference in priority attributed to this item between F and P...

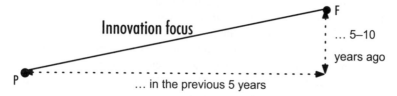

Figure 8.2 Explanation of Figure 8.1

Table 8.1 and Figure 8.1 indicate that front-runners have put great emphasis on implementing new information systems over the previous five years, far more than they did five to ten years before. Another item currently high on their agenda is 'innovation focus'. Not only have they paid more attention to product and process innovation in the previous five years than in the five years before that, but they also paid significantly more attention than pack members to innovation. Currently, front-runners do not differ from pack members in the attention given to customisation, but five to ten years ago front-runners gave significantly more attention to it than pack members. This supports our findings in Chapter 6 to the effect that front-runners are perceiving the inherent limits of customisation. The data also indicate that front-runners were also much earlier with streamlining than pack members. Streamlining was already an important policy issue for front-runners five to ten years ago. At that time, cost reductions, reorganisation, concentration on core activities and implementing quality assurance were high on their policy agendas. Pack members gave significantly less attention to these issues. Not shown in the figure is the fact that front-runners attribute higher value to vocational training and hiring highly skilled people (which can be seen as actions related to organisational innovation) than do pack members.

While the attention devoted by front-runners to market expansion has slightly decreased between the two periods, the attention devoted to it by pack members has increased by 0.5 on a five-point scale over the past five years, indicating a possible timing difference. For pack members, market expansion became the most important business policy item in the previous five years, followed by customisation, streamlining and the implementation of new information systems. Over the previous five years, that is, they have started to pay more attention to streamlining their organisations, customisation, market expansion and implementing new information systems than they did during the five years previous to that. 'Innovation focus', however, scores lowest of all on their priority list.

Thus, the study indicates that front-runners have streamlined their organisations and expanded their markets earlier than pack members. These findings are supported by our qualitative data (for example, Box 8.2). Furthermore, front-runners focused much earlier on customisation and the implementation of new information systems. With respect to the past five years, front-runners differ significantly from pack members regarding the 'Innovation focus' factor. They pay more attention to innovation than pack members do. Summing up, we could state that timing in company policy seems to be an important predictor for innovative success.

BOX 8.2 ANTICIPATING THE FUTURE

Company 11 is a pack member in the wholesale business. 'Our company's strong point is our decentralised approach to the flow of goods, which enables us to supply our customers with exactly what they want,' the director says. 'Our major competitor doesn't give this kind of service any more. They have streamlined and automated their entire flow of goods and customers just have to make do with the assortment they provide. Of course, we have also started getting involved with things like IT and communication technology. When you see what is going on with your competitors, you can't fall behind.' A little later, the director considers the future: 'There will be no way to avoid streamlining the assortment in the future.' At company 44 (front-runner in the same business), this discussion had already taken place some years ago. There, too, it was a dilemma to decide to supply customers with everything they desired and still manage the enormous stock costs involved. In company 11, people are only recently beginning to discover that a decision must be taken regarding something which is now a competitive advantage but may become a competitive disadvantage tomorrow. The safest course seems to be to keep a close eye on one's major competitor.

The director of pack member 52 (food industry) has a very short time horizon indeed: 'I am more interested in next week's profit than in how my company will look in five years' time.'

Table 8.2 Competency-related scales for business policies

	Previous five years			Five to ten years ago		
	Average pack member	*Average front-runner*	*T-value*	*Average pack member*	*Average front-runner*	*T-value*
Organisational policy items	3.3	3.7	−2.42**	2.6	3.0	−2.02*
Reorganisations, quality assurance, vocational training, hiring highly skilled people (Cronbach's alpha = 0.6)						
Technological policy items	3.7	3.9	−1.30	2.8	3.2	−2.08*
Process innovations, implementing new information systems, cost reductions, concentrating on core activities (Cronbach's alpha = 0.6)						
Marketing policy items	3.8	3.9	−0.64	3.1	3.7	−2.82***
Product innovations, customisation, increasing sales, enlarging market share (Cronbach's alpha = 0.7)						

Notes: * $\alpha \leq 0.05$; ** $\alpha \leq 0.01$; *** $\alpha \leq 0.001$ (one-tailed); 1 = top priority; 2 = high priority; 3 = medium priority; 4 = low priority; 5 = no attention devoted to

The policy items are also used to construct three scales indicating the importance given to technology, market and organisational policy items in the two periods (Table 8.2). From this, we learn that five to ten years ago front-runners attributed significantly more importance to all three competency-related policy scales than did pack members. As far as the technological and marketing aspects are concerned, the difference in attention evident five to ten years ago has by now been greatly reduced.

Organisational aspects have apparently always received more attention in the business policies of front-runners than in those of pack members. This could indicate that the attention devoted to organisational aspects over the past decade may play an important role in explaining differences in innovative success.

The extent to which previous technological, marketing and organisational policy priorities are related to current strengths in competencies was therefore tested. Table 8.3 shows the correlations between policy priorities in the two periods and each of the three competencies as well as the dependent variable 'innovative success'. It indicates that reorganisations, quality assurance programmes, investments in vocational training and the hiring of highly skilled people (the 'organisational' policy items) in the past ten years have paid off in terms of higher innovative success and stronger technological, marketing and organisational competencies. Investments in process innovations, the implementation of new information systems, achieving cost reductions and concentrating on core activities (the 'technological' policy items) five to ten years ago appears to be paying off in stronger technological and marketing competencies today. And attention paid to product innovations, customization, increasing sales and enlarging market share five to ten years ago (the 'marketing' policy items) appears to be related to current strengths in technological and marketing competencies. Thus, Table 8.3 provides additional support to the importance of timely policies in explaining differences in innovative success, as it shows that timing not only helps to explain innovative success in general, but is also an important factor in explaining differences with respect to the three competencies.

Table 8.3 Correlations between competencies and attention devoted to competency-related business issues in the 1990s

	Technological competencies	Marketing competencies	Organisational competencies	Innovative success
Technological policy items				
past five years	—	—	—	—
five to ten years ago	0.33**	0.49**	—	0.26*
Marketing policy items				
past five years	—	0.22*	—	—
five to ten years ago	0.28*	0.37**	—	0.35**
Organisational policy items				
past five years	—	—	0.42**	0.30**
five to ten years ago	0.38**	0.25*	0.32**	0.25*

Notes: * $\alpha \le 0.05$; ** $\alpha \le 0.01$; *** $\alpha \le 0.001$ (one-tailed); The competencies are measured as the predicted values of the multiple regression equations of the regression formulas of the single-competency models. All correlations were positive, only the significant are shown.

These findings give support for Hypothesis 8.1. Apparently, attention devoted to organisational, technological and marketing issues (as they are defined in Table 8.4) in the past is an important predictor for strengths in those competencies (as calculated by the multiple regression models) today.

8.2.2   The three competencies combined

In order to test whether the competencies in combination provide a better predictor of innovative success than any of the three competencies alone (Hypothesis 4.2), a multiple regression analysis was performed with all the scales and variables of the singular models entered step-wise. The resulting equations rendered an R^2 of 0.62 (Appendix 4), which is higher than the R^2 of any of the singular models (Table 8.4). The resulting regression equation includes variables from all three competency models, thereby giving support to Hypothesis 3.1, that is, that a combination of competencies is a stronger predictor for innovative success.

Table 8.4 R^2 of the various models

	R^2
Technological competencies	0.41
Marketing competencies	0.26
Organisational competencies	0.36
All three competencies in combination	0.62

To test the effect on innovative success of the timing of company policy, a scale was constructed indicating the timing differences between front-runners and pack members with respect to organisational, technological and marketing policy items. This variable (called 'timing') was calculated by averaging the priorities given five to ten years before the interviews to the following policy items: process innovations, implementing new information systems, product innovations, cost reductions, concentrating on core activities, reorganisations, hiring highly skilled people, vocational training, increasing sales, customisation, enlarging market share, and quality assurance (the marketing, technological and organisational policy items identified in subsection 8.2.1). Cronbach's alpha of this scale is 0.8. Introducing this 'timing' variable into the model appeared to improve the predictability to 71 per cent (Table 8.5).

*Table 8.5 Multiple regression analysis towards innovative success on all
three competencies and the timing variable*

Independent variable	B	Beta	t
Relative R&D expenditures	0.0746	0.259	2.68**
Technological absorptive capacity	0.0426	0.156	1.62
Involving the customer	–0.0545	–0.202	–2.35*
Market-driven	0.0717	0.231	2.67**
Multidisciplinary project-wise approach	0.1387	0.506	5.78***
Timing of company policy (timely company policy)	0.0811	0.197	2.12*
Constant	0.3675		3.03***

Notes: $R^2 = 0.71$, $F(6,45) = 18.28$****; * $\alpha \leq 0.05$; ** $\alpha \leq 0.01$; *** $\alpha \leq 0.001$

Six factors can now be seen as influencing a firm's innovative success at a statistically significant level:

1. *The extent to which the firm's innovation processes are organised horizontally (project-wise) and utilise a multidisciplinary approach.* Firms which organise their innovation projects by bringing people from various disciplines together in a project team appear to be more successful innovators than firms using the linear, functionally orientated approach. In the regression analysis, this scale makes a high contribution to the explained variance and is the single most important predictor of innovative success.
2. *The relative extent of the R&D expenditures*, measured as a percentage of sales and compared to the industry average. In the regression analysis, this is the second most important predictor of innovative success.
3. The degree to which the firm sees itself as being *market-driven*.
4. The extent to which the firm *co-operates with and involves customers* during the various phases of innovation processes. Note that this variable appears to correlate negatively with innovative success.
5. *The firm's technological absorptive capacity*, measured as the extent to which the firm is prompt to detect technological and market developments; the extent to which the firm is regarded as prompt to adopt new technologies; the ease with which new technologies are used to reduce production costs (relative to the sector average); and the extent

(relative to its competitors) to which the firm is actively involved in the transfer of technological knowledge.

6. *The timing of company policy.* Successfully innovating firms gave issues such as product and process innovation, recruiting highly skilled people, vocational training and implementing quality assurance, high priority on their agendas earlier than less successful firms.

8.2.3   Predicting innovative success: crystal gazing?

The multiple regression comparisons were made with the constant variable 'innovative success' as dependent, and will have to prove their value in further scientific research. As far as the business relevance of this study is concerned, there was still the question from the sponsor as to whether a diagnostic instrument could be developed which, on the basis of a number of company characteristics, could indicate whether or not the company was a front-runner. As discriminant analysis allows us to distinguish between several mutually exclusive 'natural' groups which cannot be manipulated experimentally (as is the case for front-runners and pack members), it is especially suitable as a tool to answer this question. We therefore used discriminant analysis to derive the (linear) combination of predictor variables which best discriminate between front-runners and pack members. This technique selects predictor variables on the basis of their contribution to the correct classification of cases into the two predefined groups. A linear discriminant function is optimal if it minimises the probability of misclassification. The technique takes account of the interrelationships between the predictor variables. Translated into the terms of this study, we are thus looking for a combination of company characteristics which will enable us to reliably classify a firm as front-runner or pack member. If the company classifications in the sample appear to correspond reasonably well to the actual classification (as provided by the expert), we have a good model. We shall know in which respects a front-runner distinguishes itself from a pack member.

Discriminant analysis produced a linear equation[3] which correctly classified 90 per cent of all companies (Table 8.6). Except for 'innovativeness', the variables which proved to make the highest contribution to the classification of cases were the same as the ones which originated from the multiple regression analysis,[4] namely: the 'multi-disciplinary project team', 'involving the customer', 'timing of company policy' scales and the 'market-driven', 'trend-setter', 'relative R&D percentage' and 'innovativeness' variables.

Table 8.6 Results of the standardised discriminant analysis

Z = +0.880 MULTIDISC	+ 0.669 MARKTDRIV	− 0.483 COOPCUST	−0.430 TRENDSET
(0.000)	(0.000)	(0.000)	(0.000)
+0.406 RELATR&D	+ 0.405 TIMING	+ 0.244 INNOVATIVE	
(0.000)	(0.000)	(0.000)	

Where: MULTIDISC = Multidisciplinary project team
         MARKTDRIV = Market-driven
         COOPCUST = Involving the customer
         TRENDSET = Trend-setter
         RELATR&D = Relative R&D percentage
         TIMING = Timing of company policy
         INNOVATIVE = Innovativeness

Statistics:

Canonical correlation coefficient: 0.815
Own value: 1.976
Significance: 0.000

Canonical discriminant functions evaluated at group means:
Pack members (Group 1)   –1.489
Front runners (Group 2)   1.276

Correctly classified
Pack members	88%	(25 cases)
Front runners	91%	(32 cases)
All cases	90%	(57 cases)[a]

Note: [a] Six cases had missing values in one of the variables and were therefore excluded from the discriminant analysis.

The discriminant function classifies 91 per cent of the front-runners and 88 per cent of the pack members correctly (overall: 90 per cent).[5] The discriminant function provides a linear combination of the discriminating variables that maximises the distance (separation) between groups. Table 8.6 shows the results of the discriminant function. Although standardised discriminant coefficients are somewhat more comparable than their non-standardised counterparts, they cannot be used to assess the importance of the individual variables since each depends on the other variables in the equation. However, one might state that it is very likely that the 'multi-disciplinary project team' scale contributes most to the discriminant function. It is selected first and has the highest coefficient. The least important is most probably the 'innovativeness' variable. This is also the variable which failed to reach significance level in the multiple regression analysis. Based on this model, it is now possible to study a company and, on the basis of a small number of criteria, indicate whether it is a front-runner or a pack member.

8.2.4 Two out of three ain't bad

Now that support has been found for the hypothesis that the competencies in combination provide a better explanation for innovative success, the question is whether front-runners excel in all three competencies or whether a strength in only one or two competencies is sufficient. Recall that Hypothesis 4.2 predicted that front-runners will excel in all three competencies rather than in any single one. Discriminant analysis also enables us to say something about the number of competencies in which a firm should be strong. In order to be able to find support for this hypothesis, three 'discriminant' models are constructed. The competency variables can be seen as factors promoting innovative success. They are either organisational factors, technological factors or marketing factors which are highly correlated with innovative success. With this assumption, it is then possible to calculate discriminant functions for all three competencies separately. The question was then whether a front-runner had to be strong in one, two or three competencies. Thus, the second step is to calculate a separate discriminant model for each group of competency variables. These models classified the firms correctly in 75 per cent, 72 per cent and 77 per cent of the cases for the technological, marketing and organisational competencies respectively. From the fact that these submodels are less accurate in predicting 'overall' innovative success (their 'predicted-right' scores are lower than the predictability of the total model), we might conclude that front-runners do not always score high on individual competencies and that not all pack members score low on all three. Thus, success might even be achieved with a weak score on one of the three basic competencies. For each firm, the strength of the firm's position with respect to the three competencies was calculated (that is, the model predicted front-runner status).

From this exercise, it appears that there is a connection between the number of competencies in which an organisation is strong and the success of its innovative efforts. Being highly competent in all three competencies is strongly associated with being a front-runner. As Table 8.7 shows, 83 per cent of all front-runners are strong in at least two of the three competencies. In contrast, only 28 per cent of the pack members are well-endowed in two competencies, and none in three competencies. The T-test regarding the amount of competencies in which a firm is strong is highly significant. Pack members have an average of 0.9 competencies, whereas front-runners show an average of 2.4 competencies ($t = -7.21$, $p < 0.000$). Furthermore, we learn that all front-runners have at least one competency. These results indicate the possibilities of compensation. Obviously, firms do not have to excel in all three competencies to be successful innovators. Perhaps a comparative weakness in one competency can be partly offset by a strong position in other competencies.

Table 8.7 Group membership and number of competencies (%)

	Pack members	*Front runners*
No competencies	36	0
One competency	36	17
Two competencies	28	23
Three competencies	0	60

8.3   DISCUSSION: TRIAD OR DYAD?

8.3.1 Timing and integration

With respect to the timing of company policy, three findings are of particular importance:
1. Innovative success requires long-term tenacity with respect to technological, marketing and organisational aspects.
2. Front runners move with the times earlier than pack members and pay attention to a broader range of policy items.
3. There are indications that attention devoted to organisational policy items in the past decade is of particular importance in explaining differences in current innovative success.

The results thus point in the direction of proactive policy. Front-runners clearly began to restructure their organisations earlier than pack members (as Box 8.1, for instance, illustrated). Pack members were later in giving priority to organisational streamlining, customisation and the implementation of new information systems. However, it is not only a matter of doing things ahead of the competition: the main thing is to do things ahead of the competition which are hard to copy, or which in any case will take a lot of time to copy, such as the organisational rotation system we mentioned earlier. Another example is found in Hedlund (1994, p. 80), who describes the differences between internal processes of knowledge exchange in Japanese and western companies. In the West we devote attention primarily to increasing the transferability of knowledge. Western companies usually process knowledge in a highly mechanistic and explicit manner. But this also makes them easy to copy. Internal knowledge exchange processes in Japanese companies, on the other hand, appear to be difficult to imitate precisely because of their highly implicit character. More generally, Hedlund states that in organic environments the emphasis is primarily on implicit knowledge processing (which is so difficult to copy), while bureaucratic environments place much more stress on formalised explicit knowledge. An organisational development of this nature cannot be adopted between one day and the next, of course. There is thus a clear link between priorities in organisational development and strengths in organisational competencies.

But it is not only the organisation which was taken in hand and adjusted to the requirements of the 1990s. Front-runners also began to work with aspects that we classify under the marketing and technological competencies earlier than pack members. One conclusion which can be drawn from the analyses is that it is important to change course in time. Front-runners thus seem more apt than pack members to recognise that certain routines are becoming antiquated and are also quicker to develop new and better routines.

Various possible explanations can be presented for these findings, which moreover may well be simultaneously valid. Thus, it could very well be that front-runners were already innovating more successfully than pack members ten years ago and were in consequence forced to restructure their organisations and reinforce their marketing and technological competencies. Innovation implies that a firm does things that others have not yet done and which are hard to copy or imitate. It demands an active and proactive approach. Thus, companies which innovate more frequently and, above all, sooner than others in their sectors will also sooner encounter organisational problems to be overcome before the firm can innovate successfully. But once it has taken all the hurdles, the company will still be ahead of the pack. One explanation could thus be that front-runners were quicker to bring order into their own houses because their innovative efforts obliged them to.

The question of the time perspective is important. After all, our study was basically a snapshot of a given moment in time. However, it is also important to measure the effects of change over time. The analysis demonstrated that front-runners react sooner than pack members in certain areas, a finding which has also been shown in other studies. Studies by Moss Kanter (1983, 1989a, 1989c), Peters and Waterman (1982), Schreuder et al. (1991) refer only to the fact that successful companies are more action-prone and proactive than the less successful ones. The question which then arises concerns the position the company has come from. Was it a pack member? and hence, is its current position based on changes in company policy (interventions) in the recent past? Or is the front-runner just building on the leading positions it already had? The quantitative data provide no indications in this respect.

For example, while it appeared from the various qualitative interviews that the preliminary process is important, no clear pattern for this could be found. In fact, currently successful companies have very different histories. There are front-runners which have always been front-runners (companies 21 and 30, for example), but there are also front-runners which have had to pass through a period of (extreme) crisis (such as companies 7 and 54). This traumatic experience seems to have had a very therapeutic effect: once having been through a time like that, no one wants to go back. Different patterns and different lines of reasoning are thus simultaneously possible. That is what will make it so interesting to repeat this study after a number of years. The differences in reasoning, however, change nothing with regard to the observation that the preliminary process is evidently important.

There is still a third possible explanation for the difference, which is that front-runners are so successfully innovative at the present time precisely because they began sooner. Currently, pack members no longer differ from front-runners with respect to a number of points. The pack members appear to have made great progress in catching up. But the front-runners are still further ahead. 'Innovation' now has their full attention. And they differed significantly from the pack members with respect to the amount of attention given to innovation in company policy over the past five years. The remarkable thing, however, is that the differences with respect to the priority ascribed to innovation five to ten years ago were virtually nil. Did front-runners cram their organisations five to ten years ago to prepare for today's innovative mission? Perhaps that would explain why pack members are less successful in their innovative efforts than front-runners: they are still so busy preparing their organisations for innovation that innovation itself cannot yet get the attention it deserves.

This conclusion corresponds to the findings of an earlier twin study in the Netherlands into bear-market success strategies (Schreuder et al., 1991). This study showed that, when confronted with bear markets, successful firms take anticipatory measures in the areas of management, production scale and marketing. They quickly and vigorously adjust to the changing market conditions. A possible explanation for this more proactive policy could be the stronger external orientation (Chapter 6) and critical capacity found in front-runners (Chapter 9). Thus, another study (Stacey, 1993) indicates that successful companies are above all good at picking up weak signals from both the environment and their own organisations. Anyone can pick up strong signals. A second possible explanation for this more proactive policy might be found in the more horizontally developed organisation employed by front-runners. During the past five to ten years, the policy agendas of most companies have been filled to overflowing. Companies can no longer afford to tackle tidily one topic after the other and pass them on to separate departments. The only way to stay competitive is to keep a constant eye on the coherence of the organisation as a whole (Cobbenhagen et al., 1994a). This conclusion underscores the importance of integrating the three competencies and of devoting attention to the internal organisation.

The importance of giving attention to one's own organisation was further supported by the differences in the attention paid to organisational policy items by front-runners and pack members. Throughout the past ten years, front-runners have significantly and consistently paid more attention to organisational policy items than pack members. This constant difference regarding organisational policy items might be an indication of the higher value and attention given by front-runners to (re)designing their organisations, which underscores our earlier hypothesis that innovating firms usually pay more attention to marketing and technological aspects than to organisational aspects. Furthermore, the significant correlation between the priority given to organisational policy items and technological competencies once again emphasises the importance of a technology management function linking technological and organisational capabilities. Even at times when there is no direct economic need to do so, companies will have to invest not only in new products and processes but in their own organisations as well.

The study provides further indications that there is a significant difference between front-runners and pack members with respect to timing differences. Front runners adapt their organisations sooner than pack members. For example, we have seen a number of companies which have consciously streamlined their organisations in the past with the express goal of being able to innovate successfully in the future. Only when profitability had increased and the organisation was ready for innovation did they begin to focus on business development (Box 8.3). So the picture is not that black and white in the sense that the front-runner looks ahead technologically and the pack member does not.

8.3.2 An integrated view

The multiple regression models showed that no single competency alone was predominant in explaining overall innovative success. As no single competency correlated significantly more with front-runner status than any of the others, we can state that all three are almost equally important in explaining innovative success. The differences were small. Firms differed most with respect to the organisational competencies. Only 25 per cent of the pack members had strong organisational competencies, compared to 80 per cent of the front-runners ($t = -4.93$, $p < 0.000$). Next come the marketing competencies (32 per cent of the pack members and 80 per cent of the front-runners) and technological competencies (36 per cent versus 83 per cent). However, these differences are too small to enable us to draw any conclusions. We can, therefore, conclude that the evidence does not indicate that one competency explains innovative success any better than the others. In fact, these findings indicate that technology, marketing and organisation are so interwoven that they cannot be approached as separate processes, and that it is the front-runners, in particular, which are more apt to recognise the relevance of an integral view of company processes.

Thus, based on the analysis in this chapter, we can state that it is very likely that innovative success is to a large extent based on a combination of strengths in at least two of the following three competencies: technological competencies, marketing competencies and organisational competencies. This implies that a weakness in one of the competencies can be compensated for by strengths in the other two. Furthermore, it implies that the distinction between front-runners and pack members can be made at two levels: the level of the entire organisation and the level of the separate competencies. The latter level allows us to learn from the smart things done by some of the pack members.

BOX 8.3 SHORT-TERM POLICY TO SECURE THE LONG TERM

During a strategic orientation session, the top management of a major international supplier of raw materials came to three unpleasant conclusions: in the first place, that the R&D budget had been cut back too drastically; In the second place, that a considerably higher sum would be required to effect real strategic breakthroughs; and finally, that the company lacked the money to really make its presence felt. The company decided on a remarkable way out of this trap. For a period of three years, it would employ most of its capacity to effect process improvement and increase the efficiency of ongoing processes.

The expectation of resuming the 'real, important work' in three years' time had a very motivating effect on the product developers and process technologists involved. After three years, the company's cash cows were again in top shape and a new strategic R&D programme could be launched.

NOTES

1. This is relative to the year in which the majority of the interviews were conducted; thus, 'the previous five years' covers roughly the 1987–92 period and 'five to ten years ago' covers the 1982–87 period.
2. 'The previous five years' relates to the five years preceding the year in which the field research was undertaken, that is, 1987–92.
3. As the independent variables do not have a multivariate normal distribution, the use of a linear discriminant function is not optimal at first sight. However, this is an accepted practice (Lemmink, 1991). Furthermore, from other research (Moore, 1973, cited in Lemmink, 1991) we find strong indications that a linear discriminant function can also be satisfactory in cases where the variables do not have a multivariate normal distribution.
4. Differences between the variables in the multiple regression analysis and the discriminant analysis can be attributed to the difference in dependent variables. In the multiple regression analysis, the dependent variable was the constant variable 'innovative success' , while in the discriminant analysis it was the front-runner/pack-member dichotomy.
5. It is very likely that these are the upper boundaries of the percentage of cases correctly classified and that including more cases might lead to lower percentages.

PART III

Discussion

9. Locus of control

> You can only learn something if you are capable of comparing factories with offices. (Prof. dr Ulbo de Sitter)

This discussion chapter starts with a summary of the main findings of this study. I shall discuss alternative explanations to findings which differ from what was expected. Next, I shall present new insights that have emerged from our study. It is argued that front-runners are driven by an *internal* locus of control, whereas pack members are more likely to have an *external* locus of control. These insights, which are based on the quantitative data as well as our own qualitative observations, can be regarded as elements for new theory. I shall conclude this chapter with a reflection on the methodology used, the practical implications of our study as well as issues for further research that stem from our findings. Lessons will be drawn for the future strategy of our own research programme.

9.1 SUMMARY OF FINDINGS

The last chapter is an appropriate place to look back on our tracks. Why did we start this adventure and what have we encountered on our exploration? As was argued in the first two chapters, there is already a vast amount of knowledge available on managing innovations. But the current study differs from most others on two levels:

- it is a study of innovative success at the company level rather than the project level;
- it is a study of non-sector-specific innovation management success factors rather than single-sector success factors.

The study has focused on the broad range of companies between low- and high-tech in both service and manufacturing sectors. In total, 63 Dutch firms from 35 different industries and service sectors were studied. In these companies, developments generally proceed in an evolutionary rather than a revolutionary manner. Thus they seldom put really new products on the market but frequently effect (significant) changes to existing products, variations on familiar themes, repositioning or clever new applications (markets) for existing products, involving hardly any technological change. Sprucing up existing products with consumer needs in mind is never a very spectacular process, and the results are not often looked on as true innovations. However, they are usually essential developments. Small steps such as these can be very decisive for success on the market. We are here concerned with the critical success factors for innovation in this type of company.

With respect to these critical success factors, the study tries to achieve three objectives: testing hypotheses, building theory and providing knowledge for practitioners. This reflects the interests of the study's two target groups: the practitioners and the academic community. In fact, we might state that the study is built on two pillars: a practical and a scientific.

The practical pillar
In the practical sphere we want to help meet the need for studies which focus on the organisational context in which innovation processes are conducted and provide general devices and tools that managers can utilise to keep their organisations successfully innovative for an extended period of time.

The scientific pillar
Scientifically we want to construct a model explaining innovative success in a wide range of industries. Partly this was achieved by developing new theory based on case studies, and partly by testing hypotheses. In this respect, we have presented a theoretical framework in Chapter 4. It was postulated (Hypothesis 4.1) that innovative success in firms is determined by the presence of organisational, technological and marketing competencies. In addition, several other hypotheses were tested.

To achieve these objectives, we have chosen a case survey methodology executed in a matched sample setting. As the data gathering involved open interviews as well as precoded questionnaires, theory building could be based on both qualitative and quantitative data. Furthermore, as we employed a combination of various sources (internal and external to the company), we were able to triangulate the findings. This was done by checking for inconsistencies in the data and requesting clarification if inconsistencies were found. In addition, we explicitly chose to cross-validate the findings in follow-up studies (of which two are already underway). The complexity of the present study made it necessary continually to find pragmatic solutions to unforeseen problems as we went along in practice. This meant that compromises had to be made between the desirable and the feasible. That problem was recognised right from the beginning, and all such compromises are explicitly described in this book. In this respect, the present study, to borrow Donald Schön's swamp metaphor, resembles more a voyage of discovery in the swampy lowlands than a risk-free survey of reclaimed, familiar terrain.

This study originated in 1991. At that time, the notion of core competencies had only recently gained (renewed) attention in business and academic literature. We set out to operationalise these concepts. A large questionnaire was used to identify a diverse range of company characteristics, capabilities, strategies, and so on. The underlying notion was that front-runners would differ from pack members with respect to marketing, technological and organisational competencies. However, since no usable models were available in which the 'competencies' concept was made operational, one of the aims of this study was to explore ways in which these competencies can be operationalised. Partly this was done by testing several hypotheses related to critical success factors at the level of the firm based on single-sector studies to determine whether these success factors would also hold true for a non-sector-specific sample. In this regard, the competency model served particularly to enable principles of categorisation to be applied. The ultimate goal was to advance a further step in the development of the theory, and indeed to provide a contribution to a theory which endeavours to bridge the gap between resource-based theory (and in particular the 'core' competencies literature) and the innovation management literature.

At that time, the 'core competencies' resource-based theory had a strong technological focus; although this has diminished over the years, this theory is still very much dominated by technology. In their definition and further elaboration of the 'core competencies' concept, Prahalad and Hamel (1990) focus on technological competencies. Managerial, organisational and marketing competencies are barely discussed. The recent literature on innovation management, however (for example, Cooper, 1988, 1990; Souder, 1987, 1988; Clark and Wheelright, 1993), focuses particularly on the organisational aspects of innovation. Snow and Hrebiniak (1980), for instance, have identified distinctive competencies in general management, financial management and product research and development. Amit and Schoemaker (1993) identify the following non-technical capabilities: highly reliable service; repeated process or product innovations; manufacturing flexibility; short product development cycles. Among policy makers, as well, the notion has grown that for the average small and medium-sized companies (roughly 80 per cent of all European companies) access to technological knowledge and knowledge transfer forms significantly less of a bottleneck than access to organisational and marketing knowledge (for example, Cannell and Dankbaar, 1996; Cobbenhagen et al., 1996). We therefore argued that competencies and capabilities are not just technology-based but can be found in many business processes. As we explained in Chapter 2, rather than focusing on the core competencies of the organisation, we focused on the 'managerial competencies'. Managerial competencies are the capabilities for effectively co-ordinating and redeploying internal and external competencies (Teece and Pisano, 1994, p. 538). They can be regarded as the glue that keeps it all together and the catalyst for achieving competitive advantage. Managerial competencies go beyond simple management techniques and, similar to key capabilities, they should be regarded as combinations of organisation-related techniques, attitudes, working methods and so on. We set out to identify those managerial competencies which companies have built up in the past and which have enabled them to become successfully innovating companies. In addition, we are particularly interested in managerial knowledge about competencies and capabilities which is transferable to and useful for other companies (as building blocks for an action-relevant theory). A distinction was made between technological, marketing and organisational managerial competencies.

In addition to the hypotheses serving to secure knowledge for a broad range of industries (5.1–7.3), a number of hypotheses to test new ideas were also included (4.1, 4.2 and 8.1). In Table 9.1, it can be seen whether strong and significant support was found ($\sqrt{}$); some support was found ($\pm$); the data were inconclusive (0); the data indicated that the relation might be the reverse of the one hypothesised (–); or it was not possible to test the hypothesis (?).

As is evident in Table 9.1, a number of assumptions from the literature regarding critical success factors have been confirmed in this study for a broad group of sectors. It was shown, for instance, that for many sectors, both manufacturing and service, innovative success appears to be related to:

- the presence of organisational, marketing and technological competencies (Hypotheses 4.1 and 4.2);
- a flow-orientated, lateral and multidisciplinary method of organising for innovation (Hypothesis 7.1);
- an outward-looking mentality (Hypothesis 5.3);
- market leadership with respect to introducing new products (Hypothesis 6.2);
- a proactive management style (Hypotheses 8.1 and 5.3).

Table 9.1 Indication of support found for the various hypotheses

Research hypothesis		
4.1	Innovative success in firms is determined by the presence of organisational, technological and marketing competencies	√
4.2	If competencies are jointly present, firms are more likely to innovate successfully	√
Technological competencies		
5.1	Successfully innovating firms do not invest significantly more in R&D than less successfully innovating firms	–
5.2a	Successfully innovating firms are technologically more sophisticated than pack members	0
5.2b	The technological scope of successfully innovating firms is broader than that of pack members	0
5.3a	Successfully innovating firms are more externally orientated than pack members	±
5.3b	Successfully innovating firms are more proactive in tapping into externally developed knowledge than are pack members	√
Marketing competencies		
6.1	Successfully innovating companies are more customer orientated than less successfully innovating companies.	–
6.2	Successfully innovating firms have a great influence on new product development in the markets they serve and are among the first to introduce new products	√
Organisational competencies		
7.1	Companies using a project-orientated approach towards innovation are more likely to be successful innovators than companies using function-dominated structures	√
7.2a	Companies using multidisciplinary teams innovate faster than companies employing a	?

	linear approach	
7.2b	Companies using multidisciplinary teams are more successful in their innovative efforts than companies employing a linear approach	√
7.3	The capacity of a company to innovate successfully will increase to the extent that its culture can be characterised as creative, non-bureaucratic and progressive	±
Timing		
8.1	Front-runners differ from pack members in the timing of company policy with respect to the development of their competencies	√

Notes: √ = strong and significant support; ± = some support; 0 = inconclusive data; — = relation may be reverse of that hypothesised; ? = not possible to test hypothesis.

The analyses have revealed several relationships between organisational, marketing and technological abilities on the one hand and a firm's innovative success on the other (Hypothesis 4.1). In addition, it has been shown for a wide range of industries and service sectors that, as Hypothesis 4.2 postulates, a combination of organisational, technological and marketing competencies is an even better predictor of innovative success than any single competency. A multiple regression model has been constructed on the basis of these three competencies which explains 69 per cent of the variance in innovative success. Being highly competent with respect to all three competencies is strongly associated with being a front-runner. However, firms do not have to excel in all three competencies to be successful innovators: a comparative weakness in one competency can be partly offset by a strong position in other competencies. Successfully innovating companies tend to have a holistic view of business management. They tend to manage innovation in an integrated way in which competencies in one area are strengthened by competencies in another.

At the level of the competencies, several conclusions can be drawn. Both front-runners and pack members appear to have a strong market-pull orientation in product innovations. Based on the quantitative data, we cannot state conclusively that front-runners are more externally orientated than pack members with respect to product and process innovations. Only with respect to technological developments did we did find some qualitative, yet not significant, indications that front-runners are more externally orientated than pack members. The impressions we received from the interviews indicate that front-runners tend to look more outside the company than pack members. The differences in external orientation are not so much related to the outsourcing of R&D money or the attribution of more value to outside sources or external co-operation. But they do seem to be related to internal aspects such as the prompt adoption of new technologies, the capacity to detect relevant new technological developments at an early stage (before their competitors do), the ability to take the lead with cost reductions by implementing new process technologies and active involvement in technological knowledge. In fact, the analysis showed that front-runners are more proactive in tapping into externally developed knowledge than pack members (Hypothesis 5.3b). A scale measuring certain aspects of the company's technological absorptive capacity proved to be significantly correlated with innovative success.

Front-runners differ from pack members in that they regard themselves as trend-setters more than pack members do. Front-runners see themselves as prompt to seize market opportunities. They are among the first to introduce new products and most front-runners are regarded as market leaders in their main product group. They can therefore be said to have a major influence on product developments in the markets they serve (Hypothesis 6.2).

The current study strongly supports earlier studies which point to project-based (matrix) structures as the most preferred structures for achieving innovative success and indicate the positive contribution of high levels of multilateral relations (cross-functional co-operation) to achieving project success. Strong support was therefore found for Hypothesis 7.1, which asserted that companies using a more project-orientated approach towards innovation are more likely to be successful innovators than companies using functionally dominated structures. In 68 per cent of the cases, the front-runner twin uses a more project-orientated approach than its respective partner from the pack. In addition, we found that front-runners appear to make considerably more use of multidisciplinary teams for innovations than pack members (Hypothesis 7.2b), often appointing heavyweight project leaders.

Our research shows that organisational competencies are closely related to the concept of lateral organisational capabilities (the capacity to develop, transfer and utilise knowledge across organisational boundaries; Galbraith, 1994). Front-runners use these instruments considerably more often than pack members.

Partial support was found for Hypothesis 7.3, which postulated that the capacity of a company to innovate successfully will increase to the extent that its culture is characterised as being creative, non-bureaucratic and progressive. The 'creativity' scale and the 'progressive' scale appeared to correlate significantly with innovative success, but the 'non-bureaucratic' scale did not. In addition, the differences between front-runners and pack members only revealed themselves in the magnitude of the characteristics. There was no semantic differential on which the front-runners generally scored opposite to the pack members. The image of the average company which emerges is that of a results-orientated, flat and fairly informal organisation with relatively few differences in status. The average company is open to change, moderately progressive and tolerates failures, provided people learn from them.

Factors explaining differences in success were not only found in the present but also in the past. It was shown that timing in company policy seems to be an important predictor for innovative success (Hypothesis 8.1). When the companies were asked to look five to ten years back, it can be said that front-runners attributed significantly more importance to all three competency-related policy scales than did pack members. The results thus point in the direction of proactive policy. As far as the technological and marketing aspects are concerned, the difference in attention evident five to ten years ago has by now been greatly reduced. Organisational aspects, on the other hand, have apparently received consistently more attention in the business policies of front-runners than in those of pack members in the past ten years. This could indicate that the attention devoted to organisational aspects over the past decade may play an important role in explaining differences in innovative success.

With respect to the timing of company policy, three findings are of particular importance:

1. Innovative success requires long-term tenacity with respect to technological, marketing *and* organisational aspects.
2. Front-runners move with the times a few years earlier than pack members and pay attention to a broader range of policy items at the same time.
3. There are indications that the amount of attention devoted to organisational policy items in the past decade is of particular importance in explaining differences in current innovative success.

One conclusion which can be drawn from the analyses is that it is important to change course in time. Front-runners thus seem more apt than pack members to recognise that certain routines are becoming antiquated and are also quicker to develop new and better routines. Innovation implies that a firm does something that others have not yet done. It demands an active and proactive approach. One explanation could thus be that front-runners were quicker to bring order into their own houses because their innovative efforts oblige them to do so.

We did not find conclusive support for other hypotheses derived from single-sector studies. Contrary to expectations, for instance, we could not find support for Hypotheses 5.2a, 5.2b, 6.1 and 7.2. From the study it becomes clear that, for a wide range of industries and service sectors, technological sophistication (being the advancement of the technologies mastered as compared to the advancement of their national and international competitors) can hardly be regarded as a distinguishing characteristic between front-runners and pack members. Hypothesis 5.2a could therefore not be supported. In addition, we also failed to find a statistically significant correlation between the breadth of technological scope and innovative success, as Hypothesis 5.2b postulates. We did find some indications for this relationship, but they were not statistically significant.

Nevertheless, a scale measuring technological leadership (constructed from the variables indicating the firm's relative technological capabilities, its technological sophistication, the range of technologies it masters and the extent to which it can be regarded as a trend-setter) did show a significant correlation with innovative success. We may thus assume that it is the *combination* of various technological capabilities (for example, breadth and sophistication) which allows front-runners to differentiate themselves from pack members. In addition, we have to conclude that technological advancement appears to be more closely related to innovativeness than to innovative success, as many of the variables used to measure technological competencies appeared to be more strongly correlated with innovativeness than with innovative success. What emerges rather strikingly from our interviews is that the small and medium-sized firms are not all that involved with R&D. Much more important than R&D figures themselves is a firm's ability to pick up new developments at the right time. One could call this the antenna function. It has to do primarily with picking up weak signals; knowing whether one is more or less involved with R&D than one's competitors; knowing where knowledge can be acquired; knowing what the competition is up to, and so on. To that extent we found that front-runners were more proactive in tapping into externally developed knowledge than pack members.

Furthermore, and contrary to expectations (even for a non-sector-specific sample), we could not prove that companies using multidisciplinary teams innovate faster than companies employing a linear approach. However, on closer examination of the differences, a third variable appeared to be involved: the complexity of the innovation. Closer examination revealed that companies using a multidisciplinary approach (most often the front-runners) were generally engaged in more complex innovations than those using a linear approach. The types of problems faced during the innovation process thus differed widely. Therefore, it is not possible to conclude anything regarding Hypothesis 7.2a.

Finally, no support was found for Hypothesis 6.1, which postulated that front-runners are more customer-orientated than pack members. For both front-runners and pack members, by far the most important sources for new product ideas are those close to the market, such as customers and the marketing department. It appears that customer orientation is no longer a factor discriminating between front-runners and pack members. On closer examination of the data, we learned that the story is more complicated than *the extent* of customer orientation. In Chapter 6 we introduced the customisation paradox: customisation is not only a route to success, but can become a route to failure as well. The crucial challenge in this respect is to find a balance between the growing diversity of customer needs (external variation) and the inevitable need to simplify internal processes (internal variation). Front-runners are more aware of the limits to the diversity of their assortment and they appear to be more capable of finding this balance by setting boundaries to the range of variations. Our research therefore indicates that the relationship between customer orientation and innovative success can be better represented by an inverse U-shape (see Figure 6.3). Customer orientation contributes to innovative success only up to a certain point. Beyond that point, it can even become a hindrance to success. To front-runners, customisation is not only a process of listening to but also of convincing the client. What customers really need is as important as or even more important than what they demand. The supplier becomes the trouble-shooter. Thus, *the question as to where adapting to clients ends and educating them begins is of current interest.*

The contribution of this study to the formulation of theory should particularly be sought in the detection and identification of non-sector-specific factors explaining differences in innovative success. The study has shown that managerial competencies are involved in both service and manufacturing sectors. Moreover, the constructed model contributes to a better understanding of the integral character of critical success factors. Innovative success is not the result of doing one or a few things extremely well, but rather of doing many interrelated things well. It thus appears that front-runners not only differ from pack members in terms of technological competencies but also and primarily in terms of organisational and marketing competencies. It has also been shown that front-runners are strong in at least two of these three competencies, while pack members are at best strong in one competency.

One of the most significant findings of this study may well be the observation that innovative success is not the result of a small number of separate routines but of the system formed by these routines: the configuration. The support found for Hypotheses 4.1 and 4.2 confirms that we are on the right track. The limitation, however, is that we have had to construct this concept from the critical success factors. We have not been working on the basis of a comprehensive whole and thus cannot describe the system in this manner. Follow-up studies will have to go more deeply into this. What we can do now is indicate a general direction for this configuration. By utilising the trawler approach to data gathering (Daft and Lewin, 1993, also see subsection 3.2.1), the study allowed us to look at organisations from a broader perspective. Companies were examined from various angles (both inside and outside the company) and with different points of view (a multidisciplinary team). In total, around 285 interviews (each lasting roughly one and a half hours) were conducted. Based on this vast amount of qualitative information, one dominant theme emerges: *Front-runners are more prone to take control of their destinies themselves.* The next section will discuss these findings in detail.

9.2   AN INTERNAL LOCUS OF CONTROL

Life in the Formula 1 is very unsure. That's why you have to be sure of yourself (Mario Illien, chief designer of the McLaren Mercedes V10 engine with which Mika Hakinnen became Formula 1 World Champion in 1998 and 1999 (*F1 Racing*, August 1998, p. 43)

The sample includes eight companies which went through major crises in the recent or not-so-recent past. For example, some of them were created by the breakup of large conglomerates and concerns in the early 1980s. They became independent, were purchased by others or effected a buy-out. The extent to which this past determines the present focus is remarkable. No one desires to return to the old situation. Communication lines are kept short, business is carried out in the closest possible relation to the process, cash flows are strictly monitored, risks are spread and the market is alertly followed. 'We have learned our lesson and never want to land in such a situation again.' They are taking control.

This taking control is something we observed not only in these eight companies; it is an attitude we frequently encountered in front-runners in general. We sensed that a number of them have a quality not easily captured in a familiar organisational characteristic or cultural dimension. It is a sort of very profound basic attitude which transcends concrete management routines. This 'company quality' can perhaps best be described on the basis of the 'locus of control' concept in psychology. Rotter (1966) defines locus of control as a personality variable that describes how individuals generally attribute the cause of events in their lives. People who are more internally orientated see themselves as being responsible for events that occur. Those who are externally orientated attribute the cause of events to luck, change, fate or powerful others. In other words, people with an internal locus of control generally think that things happen because of their own choices and actions; people with an external locus of control, on the other hand, generally consider factors outside themselves to be the reasons for the events which happen in their lives. This locus of control concept could be applied metaphorically at the level of a company as well. The psychological locus of control concept at individual level, with all the theories which support it, cannot in any case be translated just like that to company level, nor is it our intention to do so. It is thus important to stress that the locus of control concept is here used as a metaphor, as a different perspective from which to examine the organisation, with the aim of obtaining more insight. This is a much-used and accepted manner of modelling in organisational theory. Although metaphors help us to see and understand organisations only partially, they are very useful in visualising and highlighting certain aspects of a problem (Morgan, 1986, p. 12).

If we apply this locus of control concept to our sample, we can state that front-runners in particular have an internal locus of control while pack members tend to have an external locus of control. Front-runners are more proactive. For example, they select the environment in which they can best flourish or try to adapt the environment. Pack members, on the other hand, are much more reactive. They attempt to adapt as effectively as possible to the environment. Causes of errors, for example, are sought primarily externally.

Applied at company level, the internal locus of control concept appears to be closely related to the resource-based view (see Chapter 2), in particular where ideas about strategy as stretch and leverage are concerned (Gorter, 1994, p. 124). This strategy as stretch and leverage paradigm starts from the premise that organisations can choose their environment (by changing their market focus, relocating production facilities, and so on) or even change their environment (by lobbying policy makers, advertising and so on). Organisational capabilities and competencies can play a major role in changing the environment. Proponents of this view (Hamel and Prahalad, 1989) argue that the foundation for a firm's strategy and success is provided by its resources and capabilities and is not determined by the available space in the industrial sector. They argue that companies which base their policy on market and competitor analysis will not differ much from each other in the end, and consequently will become engaged in a fierce competitive battle or risk lagging behind. Here a parallel can be drawn with the ideas related to the internal locus of control. In contrast to this 'inside-out' approach is the 'outside-in' approach to (innovation) strategy formation. This paradigm, also called the strategy fit and allocation paradigm (discussed in Chapter 2), is fairly deterministic in the sense that it is based on the notion that a company should adapt its strategy to circumstances. The underlying idea is that a company should choose its competitive positioning on the basis of a fit between the strengths and weaknesses of the company on the one hand and the chances and opportunities in its environment on the other.

'*Le futur n'est pas un endroit ou nous allons, mais quelque chose que nous pouvons créer*' (The future is not somewhere we are heading for, but something we can create). This statement, which is attributed to Jeanne d'Arc, is after more than 550 years still very apt. In essence, it characterises the driving force of companies with an internal locus of control. They believe they can control their own destiny to a great extent and they exert this control. In their view, the only way to outstrip the pack is by taking control of their own fate. However, it is necessary for the firm to have built up competencies which will really enable it to take control successfully. The interviews give us the definite impression that front-runners in general tend to follow the inside-out resource-based view rather than the outside-in approach. Support for this impression was found in various quantitative findings. An example: pack members and front-runners had hardly any differences of opinion when it came to characterising the dynamics of the environment: both found it to be dynamic. But where predictability is concerned, pack members tend to see the environment primarily as unpredictable, while front-runners are considerably more often of the opinion that it is predictable. Furthermore, they believed that they could exercise substantial influence on their environment. Although every strategy represents a healthy balance between ambition and possibilities in practice, the dominant theme for the front-runners is primarily internal strength: 'this is what we want to achieve in the future and that's what we're going to do'. The pack members, on the other hand, tended to view their future as primarily determined by the degree of freedom allowed them by the environment (markets, competitors, customers, government, and so on). They had a stronger feeling of being at the mercy of the caprices of the environment (Box 9.1). We also encounter these classic ideas in Porter's outside-in approach. We can thus state that there are indications that front-runners primarily have an internal locus of control and pack members an external locus of control.

We did come across *one* pack member in the sample which also had a strong internal locus of control. Since our study, this pack member has developed tremendously. In the second half of the 1990s, it is even viewed by industry watchers as a definite front-runner (Box 9.2). This company illustrates that a strong locus of control can be a means for a pack member to sprint ahead, but it also demonstrates that an internal locus of control is not sufficient in itself. One must be capable of putting it into practice. The company must have built up the competencies which really enable it to take control successfully.

BOX 9.1 BELIEF IN ONE'S OWN CAPACITIES

We are incapable of maintaining the same list of priorities for three months.' This statement by a marketing manager of a pack member (company 33) indicates how little the firm works the market on the basis of its own strength and conviction. The company predominantly reacts (*ad hoc*) to market signals.

The marketing manager of a front-runner in the food sector (company 25) has an entirely different outlook: 'It is not the consumers who come up with new ideas. That's something you generally have to take care of yourself. But consumers can often help improve products (in terms of colour, taste).'

Company 54 (front-runner) believes in its own capabilities (an expression of an internal locus of control). One could sense this immediately in the very first interview. This company did not allow ISO certification to be imposed on it by the market because it was convinced that it could serve its customers better by following its own philosophy. In the early 1990s, the market began to demand quality certification. Following careful analysis of the ISO standards, the company decided to develop its quality management system on the basis of its own standards rather than the ISO standards. In this way, the company would be better able to maintain its flexibility and adapt its quality planning to its customers. The R&D department then drafted a quality plan which can be seen as a practical application of the ISO standards. A conscious choice was made to entrust the introduction of the quality plans to the R&D department (rather than appointing a new quality manager) because this department has close ties to both contractors and customers. Although its customers more and more frequently invite tenders only from ISO-certified companies, the firm seems to occupy a unique position. The month before our visit, they even managed to take away a major customer from an ISO-certified competitor.

The pack member in the food sector (company 52) clearly had a fatalistic attitude. For this firm, innovation primarily implied process

---

innovation, in particular process innovation aimed at lowering the cost price. The director even refers to this as 'a matter of life and death' for the company. In his view, product innovations make little sense. They tried a couple of times, but the products were never very successful. When it comes to product innovations they have to compete against two large conglomerates with huge R&D budgets. From their point of view, any attempt at product innovation will cause problems right from the start.

A number of years ago, the company came up with what it considered to be a unique product, for which it would even have been able to obtain a world-wide patent. It was then decided not to invest in the product, because there was no expectation that the company could introduce it successfully and win the battle against its competitors. The result is that a similar product (prepared in a different manner) has now been patented by the firm's biggest competitor. 'So why should we go to all that trouble, we're just too small to innovate', sighs the director.

Even a superficial analysis of the failure of the firm's product innovations, however, shows that it is not the environment which is to blame but errors in the innovation process made by the firm itself (marketing blunders in particular). Organisationally, as well, the company's house has not been in order. When asked how product innovation proceeded in the company, the director replied: 'The commercial department initiates the process and from then on it's trial and error'.

BOX 9.2 A PACK MEMBER WITH AN INTERNAL LOCUS OF CONTROL

How important an internal locus of control can be to a pack member making its way to the top was shown by pack member 20 (software development). Some two years after our visit to this company, it was evident from articles in the business literature that it had developed tremendously since that time. The firm is fairly young and has grown rapidly. Four years after our visit it can be observed that this company has become a front-runner.

During our study, we were particularly struck by the company's self-confidence, which approached arrogance. It is a very technologically driven company which develops new products to a large extent on the basis of its own convictions. 'The market is not yet ripe for our philosophy, but that won't bring us to our knees. We know we will be proved right.', said the marketing manager. And he was proved right. Between 1993 and 1997, sales rose from 180 million to over 1.4 billion guilders and net income grew from a loss of 7 million to a profit of 83 million guilders. In retrospect, the fact that they stuck to their own developmental philosophy is indeed the most important pillar underpinning their success.

In the second half of 1998, however, the company encountered problems again. Apparently, these are due particularly to the fact that the internal organisation is not in order. Because of organisational problems, the company is not capable of keeping its promises to customers. An internal locus of control is thus necessary, but one must be able to put it into practice. The company must have the organisational, technological and/or marketing competencies which will enable it to be truly successful.

The internal locus of control is particularly expressed in active management of relations with the environment, a strong knowledge base and the awareness that innovation is manageable and that innovative success to a certain extent can be produced on demand (Figure 9.1). Each of these building blocks will be discussed in more detail in the following paragraphs.

9.2.1 Managing relations with the environment

Selecting and influencing the environment
The idea that an organisation and its environment are interdependent first came into vogue with contingency theories in the late 1960s. These theories suggested that various components of an organisation, most notably its structure, needed to be aligned with the demands of its business environment (Lawrence and Lorsch, 1967). Population ecologists take this contingency viewpoint a step further. These theorists adopt the biological theory of natural selection to explain that organisations which are more fit for their business environment are selected for survival, while less fit organisations are left to their demise (Hannan and Freeman, 1990). The 'fit' with the

environment was traditionally considered as an objective fact. Companies were told they would have to adapt to the environment in order to survive. This kind of Darwinian outlook, however, blinds us to a range of other possibilities. The environment is not really an established fact but can to a certain extent be selected and influenced by the company. Companies lobby the government, try to change public opinion with a variety of marketing-mix instruments, make agreements with competitors, customers and unions. Companies can also select their environment by concentrating on niches or specifically trying to penetrate (or ignore) certain markets. It makes sense for companies to exploit technological advances in the market. This demands an active attitude, not only towards customers but also towards the government, the sector organisation and other institutions which play a significant regulatory role. Front-runners appear comparatively better linked with customers and suppliers, from whom they obtain inputs for new products and processes. In other words, they are better endowed with social capital (Burt, 1989).

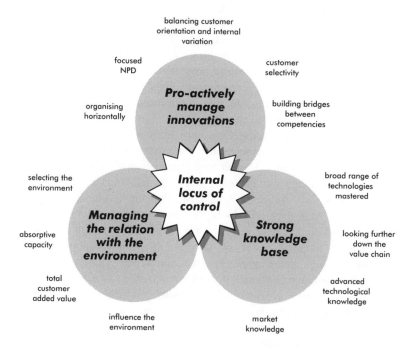

Figure 9.1 Building blocks of an internal locus of control

But communication with the environment also plays an important role. Active communication with the environment can influence the way the environment perceives the company. Thus, history teaches that a very defensive reaction to a disaster may often be counterproductive, creating an atmosphere of distrust in the public (which is part of the environment). Front-runners are much more aware that the environment can be manipulated to a certain extent. Many front-runners benefit when standards imposed from outside become more strict, for example standards related to the environment, quality, working conditions and health. They have so much technological know-how available that others will find it difficult to follow them. They therefore try to steer government policy or sector research in a certain direction. What seems to be altruistic behaviour (giving staff members free time to work for the sector) thus appears in a different light. It could be posited that front-runners generally have a more proactive attitude towards their environment, while pack members primarily tend to react.

Absorptive capacity

The qualitative data give the impression that front-runners can be said to spend their R&D money more effectively than pack members. The front-runners include companies that can pick up new technological developments from other application areas and translate them into their own products, services and processes. Their know-how is often obtained from sources outside the company. But of crucial importance here is the capacity to internalise all these external developments. It is necessary to absorb this external knowledge internally. In this connection, Cohen and Levinthal (1990) speak of the absorptive capacity.

A new hypothesis emerging from this study is that having a strong internal locus of control enables the company to internalise external knowledge more effectively. Having an internal locus of control means that the required knowledge can be more precisely defined, the company can make more conscious choices. They know what they want and how they want to get it. And they will look only for knowledge which actually corresponds with their objectives (Box 9.3). Contrast this with a company with a strong external locus of control. It is stimulated to effect improvements and innovations by the numerous signals from the environment it perceives, to which it usually reacts in an *ad hoc* manner. The question then, however, is whether the company can manage this externally acquired knowledge internally.

Total customer added value

Total customer added value is concerned with providing the customer with a complete package to solve his or her problems. It goes beyond the notion of selling a product. In Chapter 6 we have already mentioned the interweaving of the physical product and added service. Increasingly, 'products' become more than physical products alone. With slogans such as 'Our products are solutions to customer problems', some front-runners present themselves very distinctly as problem-solvers rather than mere manufacturers of products. Others don't go so far, limiting themselves to providing advice and information. None the less, they manage to represent clear added value for their customers by doing so. Such companies attach great value to building up partner relationships with customers and taking their problems into account. For some front-runners, service has evolved into a concept in which the entire organisation (processes, structures and culture) is organised around giving consumers what they want in terms of both expressed and latent needs.

BOX 9.3 LOOKING FOR KNOWLEDGE THAT CORRESPONDS WITH THE OBJECTIVES

The front-runner in the food sector makes very conscious market choices based on a good understanding of internal strengths and weaknesses. Where this company is concerned, sector experts and even competitors are clearly in agreement: it is a true front-runner. Yet the company's internal technological strengths lie not so much in the area of rapid introduction of new products as in high quality and low costs in the production process. The emphasis is thus on process innovation. The company has decided on bulk production, as it were. In addition, it has long-term contracts with a large number of smaller firms which produce the 'specialities' (the more exotic or exclusive assortment sold at higher prices) on assignment from the front-runner and under its quality control. In this way, the company ensures for itself a strong position with respect to both the cash cows and the new products.

Until recently, customising particularly implied satisfying customer needs in terms of product specifications, order size, delivery times and quality. Firms that were quick to take this direction are now carefully retracing their steps. Customising is now translated into the advantages that can be offered to the client by using the firm's own solutions. In that sense, customising involves not only listening to but also convincing the client. What clients really need is as important as or even more important than what they demand. Being 'smarter' than the client is considered an inherent feature of professional services. That is why knowledge of customer processes is so important. The supplier becomes the problem-solver.

The increasing interweaving of manufacturing production and service is a trend found in many sectors, albeit to different degrees. Even some pack members show developments in this area. This development has significant organisational consequences. Products and services are increasingly supplied by the same outlet and the internal chain is becoming more highly integrated. Companies that want to innovate in a more customer-orientated way (often by developing specialities) could almost be described as passing through an evolutionary process. They first go through a phase of divergence, in which many new products and variations are developed and put on the market. After some time, they learn from their market experiences and the assortment is trimmed and streamlined. In this convergence phase, the company decides on a definite strategy which provides direction for further innovations. A number of front-runners currently find themselves in the phase we call 'service on core product' (see Chapter 6). In this phase, companies distinguish themselves by service related to their core products. They sometimes promote themselves as problem-solvers, rather than simply producers of goods, by offering an integrated package of products and services. For these companies, service related to the physical product has become an important competitive weapon. And some have made themselves virtually indispensable to their customers through their additional service. This phase model has emerged as a new hypothesis from our study (see Chapter 6). Its validation will demand further research.

9.2.2 Strong knowledge base

The classic distinction between technology-push and market-pull, inside-out and outside-in, is too simplistic. Some of the best-performing front-runners build their innovation and marketing efforts on their own technological, managerial and marketing strengths and do not unthinkingly follow the market. Nevertheless, they seem better at knowing what the customer wants than the 'classic' market-orientated companies. The concept of a strong knowledge base thus concerns both a strong technological knowledge base and a strong marketing knowledge base.

A strong technological knowledge base can be characterised by the range of technologies mastered, the level of development of these technologies, and the company's absorptive capacity, its ability to detect and internalise external knowledge. Front-runners generally appear to be leaders in terms of technology and usually master a broader range of technologies than their competitors. The qualitative data give the impression that front-runners engage more often in tapping into other firms' know-how. The front-runners include companies that can pick up new technological developments from other application areas and translate them into their own products, services and processes. Their know-how is often obtained from afield.

Among other things, the marketing knowledge base is concerned with the company's market knowledge and ability to look further down the value added chain. Knowing and sensing changes in the consumer market provides insights in the expected changes in the business to business markets. Furthermore, we visited several front-runners who have adopted an assertive attitude towards their customers. They have realised that just listening to customers is not enough; in some cases, they have to convince customers that the company's service or product is what they really need. They have accurate knowledge of their customers' real problems and are able to provide solutions.

9.2.3 Proactively managing innovations

Innovation and uncertainty go hand in hand; many a new product is due to chance. But this does not imply that innovation is unmanageable. Far from it. For example, conditions can be created in which innovations can flourish. And despite the many uncertainties and risks, the innovation process itself is to a certain degree also manageable. However, it requires a different management style and organisation than is used in steady-state processes. Companies have to recognise this and adapt their routines.

Organisations learn when solutions based on experience can be applied to manage problems more effectively. This learning behaviour assumes a relationship between solutions and problems or, to put it differently, between cause and effect. Such a relationship seems self-evident, but it is not always so obvious in practice. For instance, those concerned with solutions (product and process design staff, for example) may be working at too great a distance from those who are actually facing the problems (production, sales and service staff). Innovation affords the greatest opportunity for learning when designers and decision makers involved with the initial stages of the process (upstream) are rapidly confronted with the (possible) effects of their choices later in the process (downstream). It is also true with respect to innovation that the control cycle must be secured with effective feedback and advance co-ordination. An important condition for this is to have a strong, horizontal, flow-orientated organisation which can tackle innovation with an holistic approach.

Along with the integration of technological, marketing and organisational competencies, this holistic approach also finds expression in the integration of internal and external knowledge development and acquisition; in a horizontal, process-based manner of organising innovation; and in the integration of products and services into total customer added value. It emerges, as well, in the integration of internal and external knowledge; the awareness that success is primarily the result of strengths in at least two competencies; emphasis on total customer added value; and a horizontal, flow-orientated organisation.

Organising horizontally

Modern management theory (Skinner, 1974; Sitter et al., 1997) points to the importance of simple organisational configurations with a clear focus. Such simplicity and focus can be achieved through flow-orientated (horizontal) organisation: by designing organisations so that homogeneous flows emerge with maximum internal interactions and minimal interactions between them. The great value of such approaches is that they point to the importance of integration in chains and of allocating control tasks as low as possible in the organisation. 'Integral design' and 'organisational learning' are key concepts in this respect.

The current study supports these ideas. It has shown that innovative success is developed best in highly flow-orientated organisations. When it comes to the innovation of products, services and processes, the most important organisational task is to break through compartmentalisation. Most companies have a past characterised by vertical compartmentalisation into jobs and professional fields, a heavy hierarchy and elaborate rules and procedures. In moving away from this 'machine bureaucracy' model they make a 90-degree turn: processes are managed along horizontal rather than vertical lines. One could refer to it as a flow-orientated form of organisation along the innovation chain. It is a product- and market-based rather than a functional form of organisation.

Front-runners have made more progress on this path than pack members. The front-runners have shown us that innovation calls for a holistic, integrated, flow-orientated management style in which opportunities are seized and fresh ideas are proactively managed into hard cash. They are more capable of managing the uncertain and stochastic process we call innovation than pack members, who tend to have a more resigned attitude to management possibilities. As a pack member said 'To a large extent it is all about luck. Either we have it and the idea becomes a success, or we don't have it and face another failure. But at least we have tried.'

Front-runners differ from pack members in that they focus on projects, while pack members concentrate on the functional organisation (Box 9.4). Front-runners tend to have overlapping responsibilities and cohesive top management teams. In addition, they tend to have comparatively dense internal networks, as evidenced by heavyweight project managers, interdepartmental career paths, and project organisations. Furthermore, front-runners are more likely to have project structures in which project managers have responsibilities spanning all activities from development to sales. A company needs such a structure in order to be truly capable of taking its destiny into its own hands. By breaking functional silos and organising horizontally it is possible to obtain an integrated view of the problems and solutions. In addition, group cohesion and identification with the innovative process (since one is involved from the very beginning) are powerful stimulants. Team members feel committed and a 'can do' mentality often emerges.

The pack member presented in Box 9.2 illustrates the importance of organising horizontally in order to reap the benefits of an internal locus of control. This firm fails to translate internal locus of control into hard cash because it lacks sufficient horizontal organisation. Companies must thus possess the organisational competencies which effectively enable them to take control successfully.

It is striking that many pack members pay little or no attention to the organisation of the innovation process, in glaring contrast to the attention given by the same companies to organising routine processes. Of course, organisational structures and management techniques will gradually evolve into structures more supportive of innovation, but this process is too slow. Remarkably, though, these companies do pay attention to the organisation if they are not satisfied with the effectiveness of the innovation processes, and consolidation appears to be the credo. Often in such cases, the first move is to adapt the organisation in the direction of more centralised management, more control. Front-runners are far less guilty of this.

The questions related to the timing of company policy show that throughout the past decade, front-runners have consistently put a higher emphasis on organisational policy items than pack members and this increased attention appears to correlate significantly with the innovative success they are currently experiencing. Front-runners streamlined their organisations several years before the average pack member did. Now they are reaping the benefits.

Balancing customer orientation and internal variation

In a number of industries, technology-push had a bad name for some time. The freedom of the high-tech virtuosi was limited and they were increasingly driven by market developments. The market called and the companies developed. Frequently, however, customers do not know what is possible, or they demand such specific requirements that, in extreme cases, companies in business-to-business markets are virtually obliged to develop a separate product for each customer. Meeting the needs of customers is one of the characteristics of successfully innovating companies, but it is also possible to go too far in this endeavour. In industrial markets, in particular, or in other markets where companies often have only a small number of customers, we see that companies sometimes have the tendency to react too quickly and immoderately to customer needs. Marketing puts out its feelers and overloads the internal organisation with suggestions for product modifications or improvements. Internally, one hears complaints that there is no longer any 'peace' in the organisation (companies 37, 52 and 63).

BOX 9.4 DISPARITY IN ORGANISATIONAL COMPETENCIES

Disparity in organisational competencies was well illustrated by the two firms in the electrotechnical contracting business. The pack member has a functional organisational design and compartmentalises innovation projects into segments allocated to various departments. In contrast, the front-runner is strongly project-orientated and frequently adapts hierarchical arrangements to suit the project's requirements. Multidisciplinary project teams are formed without reference to the hierarchy, so a project manager might have his or her superior as a member of the project team. These teams are brought together in a single location to encourage the exchange of information and knowledge. Recruitment of new employees is very much related to their fit with the organisational culture. Such practices are non-existent in its sector twin.

The study indicates that front-runners and pack members deal with this dilemma differently. A qualitative observation with respect to the marketing competencies concerned the management of complexity. Front-runners appear to strike a balance between the growing diversity of customer needs (external variation) and the inevitable necessity to simplify internal processes (internal variation). Front-runners set rather strict limits on the extent to which they were willing to customise or to modify new products in order to penetrate different market segments. Customers have to pay for special orders and unique specifications. Front-runners limit the degree of customisation, and sometimes their customer base as well, in a conscious and careful manner. Again, this can be characterised as proactive innovation management. Pack members, in contrast, tend to follow the old adage that 'the client is king' and will in fact redesign a product to meet the customer's needs. The customer is also king to the front-runners, but their own company is the emperor. By realising the danger of total customisation and limiting diversity, they are more capable of mastering their destiny than companies which try to serve the customer's every whim.

Focused new product development

When developing a new product or service, many companies have a definite path for the development of competencies in mind. They are well aware of their technological, organisational and commercial possibilities and limitations. Adventures outside that domain will only be launched when there are very good prospects of profit. The essential reasoning behind this is that a new development must not only fit in with the existing competencies, but should also add something: something new to the company. 'Something' which preferably the competitors do not have. Of crucial importance, however, is the manner in which this new element can be translated into a routine.

In the product development cycle, the introductory phase must be followed by a growth phase in which continuous incremental improvements enable the price to be lowered and the quality to be improved. The difference between front-runners and pack members lies not so much in philosophy as in implementation. One quite often encounters pack members that, following the introduction of a specific project for which a body of knowledge has been created, seem incapable of generating a routine on this basis. There is no attempt to use the acquired knowledge as a foundation for further practical development, with the result that the company is unable to earn back the development costs it has incurred. Front-runners, on the other hand, generally can be said to have a more focused new product development philosophy. They tend to make better use of product development by building on recently developed knowledge. Repeat orders provide the opportunity to learn to increase in scale or to establish routines.

In Box 9.5, an illustration of this is provided by the pair in the engineering systems industry. The front-runner maintains a proactive disposition in exploiting its newly acquired knowledge, while its counterpart tends to miss the advantage offered by recently developed technology. Such divergent behaviours highlight the extent to which a firm draws on past know-how to undertake new activities within the limits of its technological platform. The firms illustrate differences in the path-dependent quest for new or modified products.

Building bridges between competencies

In the literature, the competency concept primarily has a technological connotation. Our study shows that it is useful to split this concept into three:

besides technological competencies, companies can also possess organisational and marketing competencies. Moreover, our study indicates that success is above all the result of a combination of strengths in competencies. With strength in only a single competency, it is doubtful whether a company can become a successfully innovating company. We thus encountered a number of high-tech firms which had been highly technologically driven in the past and had 'discovered' the market in the 1980s (for example, companies 7, 39, 44, 49 and 50). They then began working on their marketing competencies. In addition, we also visited a number of companies which had developed primarily from commercial firms and had introduced technology in the company at a later stage (for example, companies 11, 14, 33, 40 and 48). They are now trying to distinguish themselves from the competition by doing clever things.

BOX 9.5 DIFFERENCES IN THE PATH-DEPENDENCY OF NEW PRODUCT DEVELOPMENT

The front-runner manufacturer of complex engineering systems typically starts by searching for a new solution and investing in development with the client (cf. 'lead-users': Urban and Hippel, 1989), but begins to explore opportunities for additional sales at a very early stage. In contrast, the pack member has a similar strategy, but follows a different implementation route. This firm has a smaller range of technological capabilities for the design of complex systems, tailored to the needs of specific clients. Rather than spotting or cultivating new clients for the same product, however, the firm moves to a different segment in order to launch re-engineered systems to fit the specific needs of new clients. The necessary re-engineering requires substantial technical resources, which are then unavailable to invest in the routinisation of the existing prototype.

We thus have two different groups: technologically driven companies which discover the market, and market-driven companies which discover technology. With this, we are on the point of entering the familiar discussion as to whether technologically driven companies should become more market-driven and vice versa. The question is not whether market-pull is better than technology-push or on which of the two the company should focus. No, the challenge is to combine both in the same organisation. The most important message is that the gap between marketing and technology must be bridged. The bridge-building can begin on either side, but the bridge itself must be sought in the organisational competencies (Box 9.6). After all, companies are accepting greater ambiguity and opting, more or less, for increasing variation. This calls for adequate organisational response. Moreover, a change in the organisational culture may be necessary. Such a transition cannot be made unless the company has the necessary organisational capacities to balance the two sometimes conflicting perspectives. This calls for a certain culture, as well as the capacities to work in a multidisciplinary, project-based manner.

BOX 9.6 THE FORMATION OF ORGANISATIONAL COMPETENCIES AS A FIRST STEP TOWARDS THE DEVELOPMENT OF MARKETING COMPETENCIES

In the Dutch insurance business, product innovations were often only variations on existing policies. But such 'products' (policies) are easy to imitate by the competition. It was thus necessary to provide added value to ensure customer loyalty. In order to obtain a stronger position in the market, company 46 (front-runner) decided on a completely new approach centred on customer profiles instead of product groups. According to company 46, when a customer chooses a company the decisive factor is the extent to which the firm supplies a number of customer values. If these values are also found in the service provided, the customer will be satisfied.

A number of years ago, the company thus carried out a market study aimed at providing insight into these customer values. The results of this study were integrated with a number of factors considered important for future development. Since then, the customer values described at the time have become an exceptionally important input and guidance mechanism for the firm's innovative efforts. The most important battle was to get the people in the organisation to accept this focus. And that was no simple matter. Thus, the change did not proceed smoothly. It became clear that the organisation did not function effectively enough and it was soon recognised that the first thing needed was to change the organisation.

In insurance companies, the insurance specialists generally took the lead. Marketeers followed. The new path the company wanted to take, however, did not fit well with such a production-dominated organisation. A change to a market-dominated organisation was necessary, which meant working in a project-based manner. To acquire this market focus, the organisation had to be turned on its head. This was accompanied by a far-reaching process of em-powerment: the traditionally separated process – thinking (staff), deciding (management) and doing (shop floor) – was incorporated into small multidisciplinary cells. Customers were then linked to these multidisciplinary teams. The most important impetus for innovation comes now from customer values and is no longer based on risk coverage. A significant expansion of the range of services has been the result. With this company, we thus see that organisational competencies first had to be built up before they were able to enter the market and acquire marketing competencies.

The pack member in the switch and installation materials industry (company 51) also had to make the transition from a production-dominated to a market-orientated company. The company originally had quite a favourable, protected position. But the opening of the European borders is confronting it with increasingly fierce competition. Its domestic market is obviously foundering and the company is seeking expansion in new markets abroad. This subjects the organisation to completely different requirements. It becomes necessary to innovate more rapidly and think in a more market-orientated manner. Not everyone in the company is convinced of this, however. Organisationally, therefore, not much has changed, beyond the establishment of a marketing department. The newly appointed marketing manager has been assigned to get the whole organisation thinking in terms of customers and the market. When asked about the company's present position and where it wanted to be in five years, however, the marketing manager could give no answer.

Being a front-runner, therefore, is not a matter of doing one or a few things right, but of doing many things right. This again illustrates the interconnectedness of critical success factors: high-level technological competencies are not in themselves sufficient to ensure success, but they must at least be accompanied by marketing or organisational competencies, or, even better, by both (Box 9.7).

BOX 9.7 A STRONG TECHNOLOGICAL BASE IS NOT ENOUGH

Company 34 (pack member) is an electrotechnical contracting firm. When the company starts to enter a new market, the first project is seen as an investment (in knowledge development, creation of name recognition, and so on) and is not necessarily expected to be profitable. Based on the knowledge, experience and contacts obtained through this first project, the company then looks for new customers for repeat orders. In the past, considerable attention was devoted to the company's *technological development,* which was seen as one of its most important competitive strengths. The body of technological knowledge built up by the company formed the basis for its endeavours to stay ahead of the competition.

Yet the company did not succeed in making the most of this technological knowledge, which even began to work against it. Customers chose the company because they knew it could deal effectively with complex challenges and was also prepared to invest in them. Recently, the company realised that it was too much under the influence of the idea that it had to develop everything itself. Or as the new director puts it: 'it's not a playground any more'. It is now recognised that, in the past, the company allowed itself to become more involved in development than was actually good for it. The company now sees the endeavour to perform in a more *market-orientated* manner as the most significant challenge it faces. This implies, however, building up an entirely different body of knowledge: marketing knowledge.

This interweaving of factors is also apparent from the statistical analyses. Frequently, variables which have no effect in isolation do have an effect in combination with other variables. The regression analysis showed that it is the combination of technological capabilities, technological breadth and technological sophistication, in addition to being a trend-setter, which allows a front-runner to differentiate itself from a pack member. This conclusion recurs in our study again and again: front-runners do many things well. That's what makes them front-runners.

Although more research is warranted, the results suggest that the competencies are cumulative in their effect on innovative success. Multiple regression models of the single competencies show little, yet significant, correlations with innovative success. In Chapter 8 it was shown that it is a combination of all three competencies in particular which accounts for innovative success.

9.3   INCREASING EXPLAINED VARIANCE

As was discussed in Chapter 3, this study is embedded in a larger research programme. It is therefore useful to record several lessons from which future research within this programme might benefit. The key concept to be discussed in this respect is the covariance. With every survey study, there is the danger that the covariances will be coloured by the manner in which the study is carried out. This holds for both high and low covariance. The latter appears to be particularly the case in this study. In retrospect, we are forced to conclude that the differences detected were not as great as expected. Many of the factors appear to have only small effects and sometimes even fail to reach significance level when other attributes are also included in the analysis. On several variables, the pack member even roughly equals the front-runner. For example, with respect to the extent to which the firm uses formal project evaluations, or the extent to which customers are used as a source of market feedback. We can list three different causes for this lack of large differences:

1. *The 'built-in' small differences in innovative success between front-runners and pack members* When cases are used to fill theoretical categories, researchers generally choose examples of polar types in which the process being studied is 'transparently observable' (Pettigrew, 1988). Sometimes this involves the extremes of the continuum. In the current research design, however, we decided not to opt for the extremes in a negative sense since this would reduce the practical relevance of the study. After all, we were concerned with revealing that 'special something' which distinguishes front-runners from average companies. Cases were chosen as examples of the categories 'front-runners' and 'pack members' and we specifically asked the experts not to select laggards. The least successful of the pair had to be a representative of the average company. Laggards would have led to more significant differences when compared to front-runners than pack members, but one can question the practical relevance of these findings for firms which are in the pack and still have the ability to change positively. This restriction

in the variance range of the dependent variable (innovative success) also implies less variation in the explanatory bases (the independent variables). It is therefore very likely that the variance in the independent variables would become larger if really poorly performing companies were included in the sample or if a random sample were drawn (since this would also increase the variation). This is already apparent when the analysis is carried out on the 20 best front-runners and the 20 worst pack members. The regression comparisons then have a considerably higher R^2. In order to test these assumptions, one of the follow-up studies will employ a random sample. In this way, we increase the chance of including laggards in the sample.

2. *The large heterogeneity of companies studied* Since we studied low-, medium- and high-tech SMEs in 35 different sectors, the heterogeneity of the sample is quite large, especially when compared to single-sector studies. In addition, the sample included both service and manufacturing sectors. However, the matched sample design was deliberately chosen to enable critical success factors to be distilled which would hold true for a broad range of sectors. By including a wide range of industries, the present study was designed to inform us about general success factors specific neither to certain strands of technology nor to specific industries or markets. The generalities common to many sectors are thus surveyed, but not the critical success factors relevant to a limited number of sectors. If more homogeneous groups are analysed, it is very likely that some differences will be greater. For example, when the manufacturing companies subset is analysed on its own, the differences between the companies become greater. One of the follow-ups to the present study will therefore focus on a single sector: the Dutch information technology industry.

3. *The study is a snapshot rather than a film* The study in its current form provides a snapshot of a company rather than a documentary film. But companies are always on the move. Firms with good names three or four years ago may have fallen by the wayside, while struggling firms may be making successful comebacks. During the study, a small number of pack members appeared to be breaking away from the rest of the field and gaining rapidly on the front-runners. We also interviewed a front-runner that was clearly losing ground to the pack member we visited. Thus, when pack members start sprinting ahead and front-runners drop behind, findings become distorted and differences will of course be lessened. Furthermore, some topics relate to practices which have become incorporated in all firms. The delta (change) often tells more than an average figure. Because of the study's cross-sectional design, however, it was difficult to detect differences in the times at which these practices

were introduced and implemented in the firms. Timing differences in company policy, however, indicate that the explanation for differences in success can be found to a significant extent in the past. If we could obtain a better grasp of the delta (change), then in all probability the differences would be greater. This will be effected in the course of a follow-up study launched in March 1998.

In conclusion, we can state that we can be fairly certain about the differences detected and that it is most probable that the differences found in other samples would only be greater.

As it is next to impossible to have an ideal sample, one can triangulate by using different sampling techniques and different samples. We have therefore decided to vary the samples and sampling methods in the next research projects within the programme. The first follow-up study is therefore a sector-specific validation study, followed by a longitudinal study. In subsection 9.5.3 the future strategy of our research programme is discussed in more detail.

9.4   PRACTICAL RELEVANCE OF FINDINGS

Research like this can prove its value in practice if it shows companies constructive actions they can undertake in order to become more successful in their innovative endeavours. Change implies a change from the present situation to a desired future situation; for example, a pack member becoming a front-runner. In this respect, the current study may be of help in indicating the characteristics of a firm in the preferred situation and in assessing the current situation. However, it does not provide direct help with respect to the process of changing from the current situation to the desired one.

9.4.1 Determining the preferred situation

The profile of a 'typical front-runner' (Box 9.8) can provide pack members with a direction for change and help them to visualise the preferred situation. This profile is related to many findings in organisational literature, but with an additional insight: namely, that a company's innovative success is founded in particular on the coherence and integration of the basic elements of company policy: technology, market and organisation.

The study can also prove its value in practice through the do's and don'ts which can be distilled from the analysis. There are a number of practical lessons which companies and policy makers could learn from this study. They are listed in Appendices 5 and 6.

BOX 9.8 THE PROFILE OF A TYPICAL FRONT-RUNNER

A typical front-runner:
- is proactive;
- does not take the environment as given;
- has adopted a flow-orientated management of innovation processes;
- manages the innovation function from an integrated and multi-disciplinary perspective;
- is more likely to exploit external know-how;
- proactively manages its external relationships;
- has a thorough understanding of the needs and wants of the value chain;
- builds collaborative relationships between functional departments;
- balances the need for customer orientation and the possibilities for managing internal variation;
- invests in training and education (both on and off the job);
- has a critical attitude, sees things in perspective.

9.4.2 Determining the starting point

Knowing where to go (that is, what the desired situation should look like) is pointless if one does not know one's starting point: the firm's current position, strengths and weaknesses (in terms of competencies). In order to change for the better, it is vital to know what has to be changed. What can be used as a base for development and where is the gap between A and B the widest? This requires insight into the current situation. Similar to a benchmark (Spendolini, 1992), the present study can be of help in assessing the current situation. In the course of this study, several models have been constructed which facilitate the formulation of a company profile with respect to each of the three competencies and a general assessment of a firm's overall innovative success. Regression and discriminant models provide reference points for firms wishing to appraise their relative organisational, marketing and technological competencies and find out whether or not they can be considered front-runners.

A spin-off from this study, an assessment tool developed on the basis of the regression and discriminant models, is now being used as a benchmark tool for companies. For each competency, the multiple regression models allow firms to calculate a grade between 0 (most unlikely to be a strong competency) and 1 (most likely to be a strong competency). Discriminant analysis has allowed us to construct a model to assess innovative success. Furthermore, the three competency models allow us to formulate competency profiles, in which a score is calculated for each competency. Of course, these scores should not be regarded as grades in a final exam, but they can be of value as an input for changes and discussion within the organisation. The profiles provide company management with stepping stones or inputs to internal discussions and may energise or give direction to efforts to change.

The value of these profiles has been indicated by the various firms which participated in this study. Each firm has received an individualised report discussing their relative strengths on each of the competencies. Although we did not attempt to measure the effect of these reports, several firms have informed us that they formed the basis for internal discussions. For pack member 27, for instance, the present study and the company profile which emerged from it became major inputs in the process of change experienced by the company from 1995 to 1998.

9.4.3 Getting from A to B

Nevertheless, knowledge of the front-runner profile and assessment of the firm's current status are not sufficient to ensure successful change. Neither answer the question of *how* to become a front-runner. This question has to do with the organisation's development strategy. Apart from triggering internal discussions (is change needed? what kind of change?) and indicating directions for change, the current study can be of limited direct help in that process of change. This is due primarily to the chosen design, which allowed for only a limited collection of longitudinal data. As we shall see in sub-section 9.5.3, a follow-up study is currently being conducted that is aimed at detecting patterns of change by studying the longitudinal data.

9.5 ISSUES FOR FURTHER RESEARCH

The last section of this book is a theoretical reflection in which we discuss possible new leads for research. First, some new hypotheses, derived from our study, are discussed. Next we make some general observations with respect to action-relevant theories on innovation management. In the final section, we discuss how these ideas will be translated in our own research programme.

9.5.1 New hypotheses

More knowledge usually leads to more questions, and this study is no exception. Questions are answered, but the study also generates a number of new questions. These questions are not stimulated by statistical relationships but by new discussions going on in a number of companies.

Further research is necessary to gauge the value of these findings. Several qualitative observations complemented the survey findings, and others provided insight into new approaches found in various (yet not the majority) of industries and service sectors which, it could be hypothesised, will most probably be related to innovative success. Most of these new hypotheses, or further refinements of research hypotheses, originating from our research have already been presented in this book. They can be seen as the building blocks for a theory of innovation management.

New hypotheses

> **Front runners use customer input and results from market analysis more thoroughly than pack members (subsection 6.3.1).**
>
> **The relation between innovative success and the degree of customer orientation in new product development has an inverse U-shape (subsection 6.3.2).**
>
> **A strong internal locus of control enables companies to internalise external knowledge more effectively (subsection 9.2.1).**
>
> **Companies wishing to innovate in a more customer-orientated manner first pass through a phase of divergence, then a phase of convergence and finally arrive in a phase characterised by service related to the core products (subsection 6.3.3).**

To these hypotheses we wish to add a fifth. It has to do with the paradox that an increasingly shorter horizon for market-derived innovations at the business unit level must be balanced by long-term technological capabilities at the corporate level. Several of the companies studied are beginning to display two tendencies in the innovation of products, services and processes: development is more closely linked to the market (a product is not developed until there is a clear market for it), and product and process developments are limited to small, incremental steps, and no investments in process technology are made until the effectiveness of the development has been proven.

Hardly any company has cause to regret the development of these tendencies, and many companies see that they still have a lot of ground to cover in this respect. The downside of this development is also evident, however. In the first place, one sees that innovation is gaining a shorter-term perspective, particularly among pack members. The focus in product development then shifts towards incremental changes and, where process technology is concerned, of waiting to see which way the wind blows and only investing in technologies which really work. After all, what happens if a competitor successfully introduces a really innovative product or a radical process innovation? The company's safe and well-maintained market niche will then threaten to change into a slide. This threat was clearly revealed in a study of large companies (Philips et al., 1994). Cutbacks in corporate research and the transfer of responsibilities to business units create the danger that the flow of strategic research will dry up and skills will be lost. The importance of technology management and continual reflection on one's own product and process-technology portfolio is becoming imperative. This message is also relevant to the medium-sized companies forming the principal part of the study population. Timing and anticipation involve not only marketing and quality homework but also technological homework. Crucial in this respect is a company's absorptive capacity (Cohen and Levinthal, 1990), which implies that external technological developments are closely monitored and adopted in time. This capacity is well-developed in many of the front-runners in this study.

It is paradoxical that, as innovation becomes more market-orientated and has closer links to the business and a short-term perspective, the importance of strategic technology management at the company level increases. The central question is not: Where do we stand tomorrow? but: Where do we stand the day after tomorrow? So an additional hypothesis might be:

New hypothesis
As innovation becomes increasingly more market-orientated, with a short-term character and thus closer links to the business, the need for technology management at the company level increases.

9.5.2 Limits to theory development

As described in Chapters 1 and 2, the aim of this study is to contribute to the development of an *action-relevant theory*, a theory providing insights with which managers can identify and which they can use as a source of inspiration for their own behaviour and decisions. 'Translatability' (Dunn and Swierczek, 1977) is thus a major issue as well. As we said previously in Chapter 2, it was not our intention to propound a new theory. The current study is one project within a larger research programme. The goal of this project was to come up with building blocks for such a theory; ideas about how to go further. We tried to secure previously developed knowledge by testing hypotheses. Critical success factors described in the literature were tested for their general (non-sector-specific) applicability. But we also added our own ideas and developed new constructs (operationalisation of the managerial competencies). As such, this study can be regarded as the fountain-head of our programme. In this section, we shall reflect on our own study. The question is: how to move on from here towards the development of an action-relevant theory on innovation management? What are the circumstances that we have to take into account? Based on what we have learned in this study, we can make a number of observations.

The goals shift
We began with a search for discriminating factors. In the process, it soon became clear that companies and the goals they set for themselves (and/or the criteria on which they focus) are constantly subject to change. That is, the goals shift. This can be a problem when conducting research into performance indicators and critical success factors. The vanguard is always doing different things. We also encountered this with the present study. Particularly in the area of marketing competencies, our precoded questionnaire appeared to be excessively based on existing, documented knowledge of the marketing discipline, knowledge which had also become common among the pack members. To distinguish itself in the market, apparently, a company must keep focusing on new goals and criteria, time and time again. Goals – and thus success factors as well – are evidently time-bound.

Remarkable in this respect was the temporary quality of certain 'strategic plans'. In some companies, the lifetime of a plan seemed virtually to be in inverse proportion to the planning time horizon. These companies face increasing difficulty in developing a strategy for the next five years and implementing it accordingly. Apparently, to them some issues were strategic on a temporary basis. It is difficult to develop a theory when the arena and the rules to play (and win) the game are constantly changing. Some companies therefore play it safe and try to innovate with the customer as closely as possible, trying to serve the customer's every whim. One can question whether these firms have an innovation strategy, apart from a strategy based on continuous interaction and adaptation. The environment is dynamic to the extent that they no longer plan, but try to adjust the organisation as well as possible to the change that can come any moment. From the theory on agile manufacturing (Goldman et al., 1995) we learn a similar message. Agility is the ability to thrive in an environment of continuous change. Some of the qualities an agile organisation must possess include the ability to realise short product cycles; to outsource and partner with other firms quickly; to excel at low-volume, high-variety production; and to have customer responsiveness pervading every aspect of the organisation (Dove, 1995). However, as we learned from our own study, the front-runners condemn unbridled customer orientation, partly because this leads to control problems within the organisation.

Things are not linear

The problem is that reality is not as linear as the linear logic (where cause and effect are closely linked) on which many of our current theories of economics and organisation are based. It is very likely therefore that there will be no one best way to organise for innovation. Some authors even state that, in these turbulent times, strategies emerge gradually and are often nothing more than retrospective rationalisations of already ongoing activities (Turner, 1996). The challenge (Stacey, 1992) is to manage the inevitable tensions between creative developments and operational imperatives in autonomous innovative groups in which feedback and advance co-ordination are distinctly non-linear. Ironically, control in these circumstances may actually mean provoking conflict around issues, encouraging divergent cultures and presenting ambiguous challenges. An organisation in a turbulent environment will find it increasingly difficult to keep on its feet with the existing structures and routines. According to Stacey, management in a turbulent environment implies giving up the pretence that the results of one's decisions and actions are predictable and calculable. The question then is to what extent can having an internal locus of control help in coping with these turbulent environments? Are companies which can be said to have an internal locus of control more capable of dealing with the non-linear reality than companies that can be said to have an external locus of control?

Companies are constantly confronted with paradoxes

Two or more forms of logic appear to apply simultaneously. According to Stacey (1993), managers need to practise two directly opposing forms of control at the same time, two contradictory forms of management which he calls 'ordinary' and 'extraordinary' management. The prescriptions of conventional wisdom may be of considerable help to companies in practising ordinary management, but they become redundant when companies are facing the tasks of extraordinary management. In the latter case, management with and of chaos becomes important. The paradox thus lies in the necessity to build on ordinary management frameworks while at the same time continually challenging and destroying them in order to create new frameworks. We face a similar dilemma in our study. We learn that the integration of competencies and coherent configurations are important to achieve success. But we also have to acknowledge that this is not enough. As stated previously, goals shift and paradoxes and misfits are part of daily practice (Box 9.9).

BOX 9.9 A LEADERSHIP PARADOX: AUTHORITARIAN AND PARTICIPATIVE

Gist-brocades also recognises the necessity of paradox management. This company participated in a different MERIT study (Hertog and Huizenga, 1997, pp. 119-126). In the late 1980s, Gist-brocades became aware that success in its most important markets was increasingly determined by market leadership and increase of scale. The Gist portfolio, however, was too broad and diverse. It was thus found necessary to dispose of a number of operations: 'growth through doing fewer things, but doing them better than the competition' was the motto. Organisationally, this new strategy translated into a far-reaching process of entrepreneurial decentralisation. Corporate R&D functions were transferred to the business units. At the same time, Gist-brocades added an important new component to the unit organisation: a technology platform to ensure long-term innovation throughout all divisions and business units.

This implies that the paradox between long and short term, between centralised and decentralised management, between incremental and radical innovation is not eliminated by simply choosing one thing or the other. Gist-brocades has become aware that the choice must include both sides of the paradox simultaneously.

Mr J. Roels, corporate director of strategy and technology, formulates this paradox as follows: 'The advantage of business units is that the joint efforts made by people in so many different capacities are directly related to the result. Results become visible more quickly, and are thus more manageable. That is the effect of choosing a market-pull approach. But each time you evaluate the business units you see that the most successful owe their success to technological breakthroughs in the past. After all, if you work with a long time horizon, you see that it is primarily the technology-driven strategies, the real innovations, which are now providing the potential for profit. For knowledge-intensive firms, the art is to

> manage the paradox between long- and short-term results' (Hertog and Huizenga, 1997, p. 120).

For example, companies can run risks by choosing for a secure road. Furthermore, we have seen that customer orientation is not only essential for innovative success but can also be disastrous. We have seen very successful companies which, perhaps somewhat arrogantly, lay down the law to their customers. They don't try to sell solutions to problems their customers don't have, but have solutions for problems that their customers aren't even aware they are facing. Or, as a company director put it: 'You have to be very good at explaining to customers what they really want.' These are all examples of paradoxes.

So, despite the striving for coherent configurations and integration of competencies, we should not be blind to misfits and paradoxes. But this makes theory building extremely difficult, because it is not easy to base a theory on paradoxes. Usually, everything fits together nicely in a theory. But such a theory in which everything fits appears to be unworkable in practice. Dealing with paradoxes also means dealing with 'both/and' choices (in addition to 'or/or' choices). Yet even with 'both/and' logic, a choice must be made eventually. It is virtually impossible to state in advance which of the two conflicting logics has to be followed (this is illustrated in Box 9.10 by a metaphor). But then falsification, one of the requirements of a theory (Doty and Glick, 1994), will be very difficult. In addition, one might expect a theory (ibid., p. 234) to be internally consistent with respect to the relationships between the constructs (variables). The problem is clear, but also modern scientists such as Stacey (1993) do not have a clear solution for it. Their ideas are very attractive and certainly not untrue, but the challenge is to build an action-relevant theory on them that can guide managers towards making their companies more successful.

Process characteristics appear to be significant

In this study, critical success factors are central. This is also in accordance with the wishes of our sponsor, the Dutch Ministry of Economic Affairs. It became apparent in the course of the study, however, that such factors provide little as bases for explanation. They are particularly useful as bases for categorisation, and have thus been presented as such in this book. Previously, we saw that all goals are time-bound and that timing plays a significant role in explaining success. Follow-up studies will have to pay much more attention to developmental processes. After all, it is precisely the process characteristics (the answers to the 'how' questions) which appear to be significant. This requires a process theory, which 'presents a series of occurrences in a sequence over time so as to explain how some phenomenon comes about' (Mohr, 1982, p. 9). We have already stressed the importance of such a process theory. In fact, the notion of locus of control is based on that perspective: some front-runners show a certain behaviour that is based on the process they went through in the past. In this section we are particularly concerned with future theory development and the ways along which this can proceed. From that perspective, we have to conclude that it will be difficult to connect the variance theory (on critical success factors) developed in this book with a process theory. But it will be worthwhile; and for more reasons than simply the triangulation of findings. However, it is not clear yet how this should be given shape. A parallel can be made with meteorology, which pre-eminently deals with dynamic systems in which millions of observations are gathered every day using the most advanced equipment. But they are still incapable of predicting the weather forecast accurately for more than three days into the future. All scientists studying company behaviour are facing a similar dilemma. It is an everlasting dilemma.

BOX 9.10 A METAPHOR

Perhaps the best illustration of this is a tightrope walker. He is able to balance, thanks to the pole he holds in his hands. But the nature of the pole determines the stability or instability of his equilibrium. Under normal circumstances his equilibrium will be stable, but if the pole is too long or too short he or his equilibrium will become unstable. The exact length of the pole that is comfortable for the tightrope walker depends on his ability, his experience and most probably also the distance between the rope and the earth.

9.5.3 Options for further study

There is as yet no theory of innovation management, nor is there any prospect of one in the near future. How could it be otherwise in a world where change has become a constant factor? Organisations on the move in turbulent and changing environments with shifting and sometimes paradoxical goals make it very difficult, if not impossible, to develop a new theory. The problem is that the borders within and between organisations seem to be constantly changing, and that classical organisational theory seems to be falling apart at the seams. This is such a major issue that we are also incapable of laying down a theory at the present time. Nor have we ever had such pretensions. At the most, we can provide a number of functional requirements (building blocks) which such a theory would have to meet. This list is to be found in Chapters 5–8 inclusive and is summarised in the first section of this chapter. In section 9.2, an attempt has been made to bring further order and coherence into this area.

The challenge is to construct a theory based on paradoxes; on 'truths' which may be simultaneously both true and false. Theory building thus acquires a different quality. It is gradually becoming clear that companies following pure equilibrium models are having difficulties. Due to the complexity and interweaving of many business processes, cause and effect relations can no longer be so easily discerned. Based on this study, we are able to give a few pointers which should be taken into consideration in formulating such a theory. To begin with, this study demonstrates the importance of configurative thinking. The analysis has identified success factors which only make a real contribution to innovative success when integrated with or accompanied by other factors. When these findings are translated into options for further study, a number of different paths will be open to us.

Describe and explain the complexity and ambiguity of reality

One could decide to create a theory which attempts to describe and explain the complexity and ambiguity of reality. This would soon become a complicated theory, however. An example of such a theory is to be found in Stacey (1990, 1992), who attempts to develop non-linear approaches to the study of complex organisational phenomena. As even quite minor changes in apparently isolated phenomena can provoke major changes in the total system, our ability to predict the course of events is limited. Stacey (1993) therefore dismisses the notion of some grand strategic plan and argues that 'ordinary management' is not enough when an organisation has to innovate and cope with the ambiguous, the uncertain and the unpredictable.

To cope with these challenges, managers must practise 'extraordinary' management, continually challenging and destroying ordinary management frameworks in order to create new ones. This implies management with and of chaos. Extraordinary management, however, is not to be practised instead of ordinary management but in addition to it: 'For the great majority of organisations, success paradoxically requires both ordinary and extraordinary management at the same time' (Stacey, 1993, p. 16). A significant disadvantage of such theories is that they are too complex to be considered true action-relevant theories. It is hard to sell these concepts to managers. Not only does it cost them a lot of effort to get to the bottom of the theory (assuming they are even willing to take the trouble), but it is even more difficult to translate these concepts into concrete action in the company (Box 9.11). It is not clear to them how these concepts can be converted into activities and conveyed to the organisation. Attempts to apply chaos theory in organisations met with the same fate. The theory was fascinating, with something mystical and challenging about it, but its suitability for translation into action was none the less very limited.

BOX 9.11 THAT'S ALL VERY WELL IN THEORY...

The manager of company 54 describes how he sees his job: 'My main task here is not to cut through Gordian knots and take decisions for people who do not want or dare to do so themselves. I have two tasks here: to keep things constantly in motion and to ensure that the compartmentalisation which crops up here, as it does in any organisation, is continually kept in check. Knowledge must flow, keep circulating. My job is to keep the pump working.'

In the subsequent interview, he gave an illustration of how difficult it is to transmit complex organisational theories and apply them in practice: 'A few weeks ago, along with colleagues from other company subsidiaries, I attended a workshop on organisation conducted by a professor of business administration. We were all supposed to fill in a questionnaire to show how our organisations worked. I didn't complete the questionnaire. I could do nothing with it. I cannot show how our organisation works in terms of structures. Naturally, I was singled out to account for my behaviour. I tried to explain how our organisation works and, in doing so, I realised how difficult it is. All you can really do is give examples of how we work based on very concrete situations. You transmit your vision by behaving consistently in a certain manner, not by delineating parameters and laying down procedures.'

Opt for a pure process approach

We could also decide on a pure process approach such as for instance, Blackler (1993) proposes. Instead of structures, he focuses on behavioural patterns. This approach, which might look esoteric to the managers in the field, certainly provides insight. One learns to comprehend things. But it is also limited in terms of translatability. The patterns are difficult to capture in theories. And here also, the result would be a complicated theory, with little application in practice. Moreover, one risks losing sight of the factors in concrete terms, making it virtually impossible to reply to questions such as: What will this mean for our R&D portfolio, training policy, and so on?

Use concepts with which managers are familiar

One could also decide to stay close to the action in the business practice and use concepts which managers are familiar with and use themselves, such as 'second wave' concepts (Toffler, 1980) or 'concepts for ordinary management' (Stacey, 1993). By staying close to the action, it can be demonstrated that these concepts are inadequate to drive the third wave of extraordinary business. This is understandable, since such theories are based closely on the current situation. The advantage is that managers can follow the underlying reasoning and relate it to their daily practice. This is primarily the path we have followed in this study.

The problem, however, is to found new concepts and connections on this basis. For example, to give real substance to a concept such as configuration. At the end of this study, it is clear that this is our problem and that it results from the design choices we made. Fortunately, we made our design so open that it was possible to establish this very definitely. It is precisely the combination of qualitative and quantitative data which enables us to look beyond and not remain caught up in a variance theory. In this respect, we based our work on a *grounded theory*. Observations which do not fit into concepts and models enabled us to work step-by-step on concepts for extraordinary management. In this process, we had no ambition to develop an entirely new 'grand theory' for extraordinary management. This was our pragmatic choice.

What choice do we make now?

We choose to follow different forms of logic ourselves by continuing on the third path but simultaneously exploring the first and second paths as well. The action-relevant theory is still our objective, which we shall continue to pursue in subsequent studies. Three principles are central in this respect:

1. considerable investment in observation in a wide range of organisations and situations (*input*);
2. using various methods (*throughput*); and
3. communicating via concepts and models which correspond well with the cognitive maps of managers (*output*), endeavouring to effect internal coherence and fit in the process.

But this is not all. We shall also have to be alert to things which cannot be captured in the concepts of ordinary management. After all, many concepts have only temporary significance. This means that subsequent studies will also have to involve longitudinal research. It is important in this respect to continue to keep up with theories covering extraordinary management and innovation processes and consider their relevance to our own work. At this stage in theory development, a battle between opposing schools would be deplorable.

9.5.4 The next steps in our research strategy

The research programme will not end with this book. The database which has been created will continue to receive input. In this way, a grounded theory will gradually be developed. The research described in this book forms, as it were, the foundation for this theory. At the present time, two studies based on this foundation are being carried out.

A sector-specific validation study
A sector-specific study with a random sample was launched first. The objective of this study is to validate the findings from our non-sector-specific study for a specific sector, in this case the Dutch IT (information technology) sector. The study is being carried out by Edward Huizenga as part of his doctoral research.

A longitudinal study for gaining insight into processes of change
The present study is subject to a significant limitation in that it was primarily a snapshot of a given moment in time. The development of the participating companies could not be clearly followed due to the lack of reliable reference points in time. Our study provides a snapshot, but organisations are in constant motion. Front-runners may fall behind, pack members may suddenly sprint ahead (Box 9.12). The change over time (the 'delta') can therefore provide additional information about these processes.

BOX 9.12 FROM PACK MEMBER TO FRONT-RUNNER AND VICE VERSA

Pack member 20, for example, developed tremendously in the years following our visit. Between 1993 and 1997 its turnover rose from 180 million guilders to more than 1.4 billion guilders and the net income from a 7 million guilder loss to an 83 million guilder profit. The company is now generally considered to be a successful innovative company.

A front-runner can also fall back to the pack. Such changes in position came up for discussion on numerous occasions during the process of selecting the companies; for instance, during our interview with the expert of the furniture industry. Based on prior knowledge, the interviewer had an idea that a certain furniture manufacturer was a definite front-runner and asked the expert

---

about this after the interview: 'That was indeed the case,' explained the expert, 'but they threw their position away. Although they are being increasingly overtaken by competitors, they are still one of the more innovative companies in the sector. The only problem is that they have been making more and more losses over the past five years. Today, the ability to design innovative furniture is no longer enough. Consumers have become more critical and have a wider range of choice, in this segment as well. The company has been slow to realise this, and is certainly no longer a front-runner.' This firm was not included in the sample.

Competencies evolve continuously. They indicate a firm's ability to acquire new knowledge and incorporate that knowledge into organisational design and in the development, production and commercialisation of new products and services. Moreover, the competencies represent more than the sum of individual skills. As was stated in Chapter 2, they can be regarded as the product of a collective 'organisational learning' process within organisations. Organisational learning capacity is difficult to determine on the basis of scores on a questionnaire. We have thus not been able to delineate these learning processes clearly. Company and even sector comparisons can provide interesting indications as to how front-runners distinguish themselves from laggards. But insight into the processes which help companies to become and remain front-runners requires more than statistical surveys. The actual learning process is difficult to infer from a questionnaire. Additional longitudinal research, combining qualitative and quantitative approaches, is needed in order to come to more solid conclusions regarding competencies as precursors of innovativeness.

In March 1998, therefore, a second follow-up study was launched during which the 63 companies from the current study are to be revisited, with attention being focused on their strategies and activities between the two visits. This allows us to detect patterns of change. In this follow-up study, company development is central. The main questions are: How do companies become and remain front-runners? Which strategic choices have been made in recent years and to what extent have they contributed to the present innovative capacity?

This book should thus be regarded as the report on a single, yet crucial, stage of our adventure through the swamp. We want to inform our base camp of what we have found so far and assure you that we are moving on. This stage is finished, but the expedition still continues.

Appendices

Appendix 1: The sectors sampled

COMPLETE PAIRS

01.2 Horticulture (ornamental)
20.6 Margarine/vegetable and animal oil and fat industry
23.1 Garment industry
25.7 Furniture industry
26.2 Paper products industry
27.1 Graphics industry
31.3 Plastic processing industry
32.5 Concrete and cement products industry
35.3 Manufacturers of machines and equipment for the food industry
35.9 Other machine and equipment industry
36.2 Switches and installation supplies
36.8 Circuit board assembly
36.9 Other electronics industry
37.6 Motorcycle/bicycle industry
52.3 Electrotechnical contracting firms
61.8 Technical wholesale business
67.3 Company catering
72.3 Goods transport
73.1 Heavy marine transport
81.1 Banking
82.2 Damage insurance companies
84.2 Accountants, bookkeepers, tax consultants
84.3 Computer services
84.3 Software firms

FRONT-RUNNERS ONLY

20.2 Dairy and milk products industry
20.7 Vegetable and fruit processing industry
21.2 Cattle feed industry
29.2 Synthetic resins, etc. industry
29.6 Medicines and bandages industry
31.3 Plastic processing industry
34.5 Metal furniture industry
36.9 Telecommunications and signalling equipment manufacturers
51.9 Fencing industry
68.2 Car repair firms

PACK MEMBERS ONLY

20.9 Cacao, chocolate and sugar processing industry
24.2 Leather goods
25.7 Furniture industry
29.5 Paint, enamel and varnish industry
32.5 Concrete and cement products industry

Appendix 2: Overview of questionnaire contents

1. **Liaison person**
   - company profile
   - organisation of innovation process
   - financing of innovation projects
   - internal and external co-operation
   - Quality management

2. **Marketing manager**
   - markets and market strategies
   - generation and detection of new ideas
   - organisation of innovation process
   - internal and external co-operation

3. **R&D manager**
   - R&D programme for innovation
   - technological position
   - organisation of innovation process
   - process innovations
   - internal and external co-operation

4. **CEO**
   - the firm's innovative efforts
   - priorities and company policy
   - company profile regarding innovation
   - internal and external co-operation
   - financial data

Questionnaire	Precoded questions	Precoded items	Open questions
1. Liaison person	17	87	19
2. Marketing manager	15	106	14
3. R&D manager	28	170	19
4. CEO	24	123	13
Total	84	486	65

Appendix 3: Persons interviewed for the exploratory case study at Chemco

RELEVANT POSITION[1]

Management
- Division Manager of the division in which FCSP is a business unit
- Head of strategical planning, Chemco
- Business unit manager FCSP (3 interviews)
- 2 project managers
- Former business unit manager of sister 'business unit'

Marketing
- Manager of 'New Developments' department (2 interviews)
- Manager of Market Development
- Market researcher

Production
- Production manager
- Plant manager

R&D
- R&D manager, central R&D
- R&D manager of sister company
- Corporate R&D manager
- Team leader R&D
- Senior researcher

Other
- Former internal consultant

NOTE

[1] The 'relevant position' is the current or past position of the respondent within FCSP or Chemco that has a clear relation to the case and which was the reason for the interview.

Appendix 4: Regression equation

Regression equation (stepwise enter) with all scales and variables used to make the submodels, excluding the 'timing of company policy' scale.

Independent variable	B	Beta	t
Relative R&D expenditures	0.838	0.283	3.02***
Technological absorptive capacity	0.0713	0.266	2.83**
Involving the customer	−0.0474	−0.178	−2.04*
Market-driven	0.0704	0.218	2.63*
Multidisciplinary project-wise approach	0.1339	0.483	5.55****
(Constant)	0.6266		29.80****

Notes: $R^2 = 0.62$, $F(5,57) = 18.28$****; * $\alpha \leq 0.05$; ** $\alpha \leq 0.01$; *** $\alpha \leq 0.001$

Appendix 5: Lessons: dos and don'ts for companies

The study can also prove its value in practice with the dos and don'ts which can be derived from the analysis. A number of 'lessons for companies' can be distilled.[1]

COHERENCE

In recent years, the policy agendas of most companies have been bursting at the seams. Companies can no longer permit themselves to tackle one item neatly after the other and allocate them to separate company departments. Rather, it is crucial to keep a constant eye on coherence within the company.

SETTING COMPANY POLICY PRIORITIES

It is very important for a management team to have a clear idea of the policy items to which they have devoted the most attention and the items which will demand the most attention in the coming three years. An interesting test in this respect is to ask each management team member to indicate the firm's present and future priorities and then to examine the correspondences and differences between them.

COMPANY POLICY TIMING

If a company wants to distinguish itself in the market, it has to start changing at an early stage. The important thing is to go on the offensive. With so many excellent products and services already being provided, it takes a great deal of effort to even begin to break away from the pack. Setting a new course, however, is a long-term process. It must be initiated at an early stage in order to profit from a head start on the competition. This demands a proactive approach throughout the organisation, which often begins by keeping well-trained employees on the lookout for developments taking place behind the scenes.

COMBINING INTERNALLY AND EXTERNALLY DEVELOPED KNOWLEDGE

While front-runners attach more importance to their own development as a source of innovation and invest more in their own development as well, they also make more use of external knowledge. The development of new technologies is a risky and expensive business and is not feasible for most companies. They have to be smart: smart enough to look for usable, already developed technologies which they can transfer to their own organisations. Front-runners look around and ask themselves constantly whether it would be more effective to develop a new technology on their own or to obtain the knowledge from others.

INSIGHT INTO RETURNS ON INNOVATION

Front-runners usually appear to have better insight into the integral costs and returns of their innovative efforts. Of course, they usually work with general figures. But the exercise of carrying out the calculations is much more important than the figures themselves. Moreover, figures acquire added value when calculated for a number of years in succession. In this way, trends can be revealed. This is important because innovative efforts usually lead to concrete financial results only after a number of years.

KNOWLEDGE OF THE VALUE-ADDED CHAIN

All the activities which the company has to keep running must constantly be examined to determine whether they correspond to market needs. This seems like common sense, but it can also turn out to be disastrous. The study makes it clear that the time of the unambiguous message is over. A competitive edge is a more complicated affair in which contradictions must be taken into account. Customer orientation, for example, can be a path to both success and downfall. This is a paradox. It is important to know when the extra turnover is no longer justified by the increasing costs of complexity. Among other things, front-runners distinguish themselves from pack members through sound knowledge of the requirements in the complete value-added chain. The customer is important, but the customer's customer is almost as important. Suppliers will only be able to sell what their customers can sell in turn.

FROM SUPPLIER TO PROBLEM-SOLVER

Front-runners are problem-solvers. They know how to make it clear to customers that their integrated packages of products and services offer enormous benefits in terms of costs and quality.

PROJECT-ORIENTATED MANAGEMENT

The most important organisational task in connection with the innovation of products, services and processes is to break through compartmentalisation. Companies have to organise themselves in a flow-orientated manner along the innovation chain, with processes being managed along horizontal rather than vertical lines. It is important in this respect that upstream units (developers and designers) become increasingly effective at assessing the consequences of design choices for people working with those choices downstream (in production or sales). With front-runners, innovation is more highly project-orientated, while pack members tend to remain within the functional parameters.

MULTIDISCIPLINARY TEAMS

With front-runners, working in multidisciplinary teams is the rule rather than the exception. All the most important links in the innovation chain should be represented in such teams. Members are chosen for what they can concretely contribute to the process rather than on the basis of their formal position.

QUALITY MANAGEMENT AS MANAGEMENT TOOL, NOT MARKETING TOY

No one can do without total quality management any longer. In this process of demonstrating quality, and thus providing security to the customer, the essential question is whether the introduction of a quality management system provides the impetus to continuous improvement or is limited to generating piles of paper to reassure customers that all is in order. It is important that the quality management system functions as a management instrument. This change demands a different style of leadership, with the vision continually being disseminated from above and new ideas constantly bubbling up from below.

ACTIVE AND EXTROVERTED

Front-runners are active in their environments. They keep their eyes and ears open and make good use of existing channels. Listening, moving around, looking around can be characterised as a basic attitude. This also holds for their responses to regulation. By staying close to the action, they remain more alert to the possible consequences of regulations on the environment, safety, quality and working conditions. They sometimes make conscious attempts to influence government policy.

BROAD ORIENTATION

A successful organisation is one which excels in a number of disciplines. It is capable of co-ordinating its marketing and technological competencies and has developed organisational routines to both streamline the primary process and promote innovation. A successfully innovating organisation is a miracle of versatility.

NOTE

1. For a more detailed list, refer to Cobbenhagen, et al., 1994.

Appendix 6: Implications for policy makers

This study also provides lessons for policy makers who are looking for programmes to increase innovative success in companies. While not all are based on scientifically proven differences, they can nevertheless be regarded as useful lessons from several of the most successfully innovating firms. Furthermore, they may be of help in directing further research on innovative success in practice.

The main policy consideration to be distilled from this study could well be the assertion that the quality of the organisational processes connected with innovation management needs more attention (Cobbenhagen et al., 1994). Studies like the present one have shown that the inability of firms to innovate effectively can often be attributed to organisational problems.

Organisational innovation as a point for attention in technological development, however, is not yet something with which most governments are familiar. Classical measures to promote innovation in small and medium-sized enterprises (an important topic on the policy maker's agenda) did not appear to be of much use in the traditional SME community. All too often, however, this was to a large extent due to the predominantly technological perspective of these government measures. The Dutch WBSO (a scheme aimed at promoting R&D) is a (recent) example of an impetus focused solely on the R&D capacities of the companies, not on their management capacities. The recently concluded MINT (Managing the Integration of New Technologies) programme of the European Commission also focuses particularly on technology transfer. The new IMT (Innovation Management Tools) programme, while a follow-up to MINT, has a much broader design, however. We also note that while the major EC technological programmes (BRITE-EURAM, for example) include a small budget for socioeconomic research, a direct connection with greater effectiveness in the innovation chain in small and medium-sized businesses is far from obvious.

Under the influence of RTP (Regional Technology Plan), RIS (Regional Innovation Strategies) and RITTS (Regional Innovation and Technology Transfer Strategies and Infrastructure) studies, the idea that the organisational aspect also requires attention is steadily gaining ground. The lesson which emerges from the present study is clear: innovation is more than technology alone and any endeavour to promote innovation in small and medium-sized firms must begin by strengthening their management and marketing skills. The development and transfer of knowledge concerning the interrelationships between organisational and technological renewal must become part of an integrated technology policy on a national or European scale. Initiatives in this area can be structured in various forms. One way would be to link technological development programmes and projects to organisational development and research, also referred to as *begleitforschung* (Cobbenhagen et al., 1994). Other possibilities include learning networks composed of companies and knowledge institutes in which experiences can be exchanged. And of course research and the development and transfer of organisational knowledge have to be supported financially.

An absolute precondition for success with such a policy is that the companies determine the agenda. It is hoped that research projects such as the present study will be able to contribute to setting that agenda. For companies certainly require organisational knowledge as much as technological knowledge to compete successfully in the market.

References

Abell, D.F. (1978). Strategic windows. *Journal of Marketing*, Vol. 42 (July), pp. 21-26.

Abernathy, W.J. and Clark, K.B. (1985). Innovation: mapping the winds of creative destruction. *Research Policy*, Vol. 14, pp. 3-22.

Abernathy, W.J. and Utterback, J. (1978). Patterns of industrial innovation. *Technology Review*, June-July, pp. 40-47.

Ackhoff, R.L. (1979). The future of operations research is past. *Journal of the Operations Research Society*, Vol. 30 (1).

Ackroyd, S. and Crowdy, P. (1990). Can culture be managed? Working with raw material. *Personnel Review*, Vol. 19 (5), pp. 3-13.

Adler, P.S. (1989). CAD/CAM: managerial challenges and research issues. *IEEE Transactions on Engineering Management*, Vol. 36 (3), pp. 202-215.

Adorno, T.W., Levinson, D.J. and Sanford, R.N. (1950). *The Authoritarian Personality*. New York: Harper.

Aiken, M. and Hage, J. (1971). The organic organization and innovation. *Sociology*, Vol. 5 (1), pp. 63-82.

Allen, T.J. (1978). *Managing the Flow of Technology*. Cambridge, Mass.: MIT Press.

Amit, R. and Schoemaker, P.J.H. (1993). Strategic assets and organizational rent. *Strategic Management Journal*, Vol. 14, pp. 33-46.

Anderson, E. and Ayers, J.H. [1992] Current topics in specialty chemicals, 1992. *Specialty Chemicals*. SRI International.

Anderson, P.F. (1981). Marketing investment analysis. In Sheth (ed.), *Research in Marketing*, Greenwich, Conn.: JAI Press; Vol. 4, pp. 1-37.

Andreasen, L.E., Coriat, B., Hertog, J.F. den and Kaplinsky, R. (1995). *Europe's Next Step*. London: Frank Cass.

Andrews, K.R. (1971). *The Concept of Corporate Strategy*. Homewood, Ill.: Irwin.

Andriessen, J.H.E. (1990). Macro en micro technology assessment geïntegreerd. *Tijdschrift voor Arbeidsvraagstukken*, Vol. 6 (3), pp. 33-44.

Angle, H.L. (1989). Psychology and organizational innovation. In A.H. van de Ven, H.L. Angle and M. Scott Poole, *Research on the Management of Innovation: The Minnesota Studies*, New York: Harper and Row; pp. 135-170.

Ansoff, H.I. (1965). *Corporate Strategy*. Harmondworth: Penguin.

Anthes, G.H. and Betts, M. (1994). R&D: measure of success. *Computerworld*, Vol. 28 (46), pp. 32-33.

Argote, L., Beckman, S.L. and Epple, D. (1990). The persistence and transfer of learning in industrial settings. *Management Science*, Vol. 36 (2), pp. 140-154.

Argyris, C. (1980). *Inner Contradictions of Rigorous Research*. New York: Academic Press.

Argyris, C. (1991). Teaching smart people how to learn. *Harvard Business Review*, Vol. 69 (3), pp. 99-109.

Argyris, C. and Schön, D.A. (1978). *Organizational Learning: A Theory of Action Perspective*. Reading, Mass.: Addison-Wesley.

Arthur, W.B. (1989). Competing technologies, increasing returns, and lock-in by historical events. *Economic Journal*, Vol. 99, pp. 116-131.

Asch, A. van and Jansen, T. (1986). *Naar een verbetering van participatiemogelijkheden van ondernemingsraden bij automatisering: Een verkennend onderzoek naar de besluitvorming bij technoligische vernieuwingen.* Zoetermeer: Ministerie van Onderwijs and Wetenschappen.

Assen, A. van (1990). Obsoletie en loopbaankenmerken van R&D-medewerkers: resultaten van een pilot studie. *Gedrag en Gezondheid*, Vol. 17 (4), pp. 162-166.

Ayres, R.U. (1969). *Technological Forecasting and Long-range Planning*. New York: McGraw-Hill.

Bantel, K.A. and Jackson, S.E. (1989). Top management and innovation in banking: does the composition of the team make a difference? *Strategic Management Journal*, Vol. 10; pp. 107-122.

Barney, J. (1986). Strategy factor market: expectation, luck and business strategy. *Management Science*, Vol. 32 (10), pp. 1231-1241.

Bauer, R.A. (ed.) (1966). *Social Indicators*. Cambridge, Mass.: MIT Press.

Beattie, C.J. and Reader, R.D. (1971). *Quantitative Management in R&D*. London: Chapman and Hall.

Beckham, J.D. (1992). Putting learning to work. *Healthcare Forum*, Vol. 35 (6), pp. 64-70.

Berendsen, H., Grip, A. de and Willems, E.J.T.A. (1991). De arbeidsmarkt voor onderzoekers 1990-2010. *Beleidstudies Technologie Economie*, Vol. 13.

Bertsch, H.B. and Stam, J.A. (1990). Naar een nieuwe stijl van productontwikkeling. In J.F. den Hertog and F.M. van Eijnatten (eds), *Management van technologische vernieuwing,* Maastricht/Assen: van Gorcum, pp. 41-61.

Biemans, W.G. (1992). *Managing Innovation within Networks*. London: Routledge.

Bilderbeek, R.H. and Hertog, P. den (1992). Position paper: *Innovatie in diensten*. Apeldoorn: TNO Beleidsstudies.

Blackler, F. (1993). Knowledge and the theory of organizations: organizations as activity systems and the reframing of management. *Journal of Management Studies*, Vol. 30 (6), pp. 863-884.

Blanck, P.D. and Turner, A.N. (1987). Gestalt research: clinical-field-research approaches to studying organizations. In J.W. Lorsch (ed.), *Handbook of Organizational Behavior*, Englewood Cliffs, N.J.: Prentice-Hall.

Blau, P.M. (1974). *On the Nature of Organizations*. New York: John Wiley.

Boccone, A. (1988). Profits evolve in specialties business. *ECN Specialty Chemicals Supplement*, January, pp. 5-11.

Bolton, J.E. (1971). Small Firms: Report of the Committee of Inquiry on Small Firms (The Bolton Report). London: HMSO.

Bolwijn, P.T. and Kumpe, T. (1989a). Manufacturing industries in the nineties. Paper presented at the second international production management conference on 'Management and New Production Systems', Fontainebleau, 13-14 March.

Bolwijn, P.T. and Kumpe, T. (1989b). Wat komt na flexibiliteit? De industrie in de jaren negentig. *M&O*, Vol. 2, pp. 91-111.

Bolwijn, P.T., Boorsma, J., Breukelen, Q.H. van, Brinkman, S. and Kumpe, T. (1986). *Flexible Manufacturing: Integrating Technological and Social Innovation*. Amsterdam: Elsevier.

Bond, R.S. and Lean, D.F. (1977). *Sales, Promotion and Product Differentiation in Two Prescipting Drug Markets*. Washington, D.C.: US Federal Trade Commission.

Booz, Allen and Hamilton (1982). *New Product Management for the 1980s*. New York: Booz, Allen and Hamilton Inc.

Booz, Allen and Hamilton (1985). Managing for the year 2000. *Outlook*, Vol. 9, pp. 12-32.

Booz, Allen and Hamilton (1991). *Integriertes Technologie und Innovations Management. Konzepte zur Stärkung der Wettbewerbskraft von high-tech Unternehmen*. Berlin: Eric Schmidt Verlag.

Bouwen, R., Letouche, J., Donckels, R. and Van Assche, K. van (1994). *Het belang van externe ondersteuning voor de groei en innovatie van KMOs*. Onderzoeksproject in opdracht van het Ministerie van Wetenschapsbeleid. Leuven: KU Leuven.

Bower, D.J. (1993). New product development in the pharmaceutical industry: pooling network resources. *Journal of Product Innovation Management*, Vol. 10 (5), pp. 367-375.

Braverman, H. (1974). *Labor and Monopoly Capital: The Degradation of Work in the Twentieth Century*. New York: Monthly Review Press.

Bridgman, P.W. (1928). *The Logic of Modern Physics*. New York: Macmillan.

Brockhoff, K. and Pearson, A. (1992). Technical and marketing aggressiveness and the effectiveness of research and development. *IEEE Transactions on Engineering Management*, Vol. 39 (4), pp. 318-324.

Brödner, P. (1990). Technocentric-anthroprocentric approaches: towards skill-based manufacturing. In M. Warner, W. Wobbe and P. Brödner (eds), *New Technology and Manufacturing Management: Strategic Choices for Flexible Production Systems*, Chichester; Sussex: John Wiley.

Brown, D. and Cobbenhagen, J. (1997) A framework for understanding the impacts of innovation management tools. In: D. Brown (ed.), *Innovation Management Tools: A Review of Selected Methodologies*. Luxembourg: Office for Official Publications of the European Communities, pp. 229-270.

Brown, D. and Cobbenhagen, J. (1998a) Innovation management in practice. In Arthur D. Little (1998) *The Innovative Company*, Cambridge, Mass.: ADL, pp. 18-46.

Brown, D. and Cobbenhagen, J. (1998b) The competence view of strategy and tools. In Arthur D. Little (1998) *The Innovative Company*. Cambridge, Mass.: ADL, pp. 10-17.

Brown, D. and Cobbenhagen, J. (1998c) Underlying capacities for innovation. In Arthur D. Little (1998) *The Innovative Company*. Cambridge, Mass.: ADL, pp. 47-62.

Bruggen, A.L.A., van der and Hertog, J.F. den (1976). Werkoverleg op afdelingsniveau. *Mens en Onderneming*, Vol. 30 (6), pp. 334-353.

Buchanan, D., Boddy, D. and McCalman, J. (1988). getting in, getting on, getting out and getting back. In A. Bryman (ed.), *Doing Research in Organisations*. London: Routledge.

Buijs, J. (1984). *Innovatie en interventie*. Deventer: Kluwer.

Buitelaar, W. and Vreeman, R. (1985). *Vakbondswerk en kwaliteit van de arbeid*. Nijmegen: SUN.

Bullock, R.J. and Tubbs, M.E. (1987). The case meta-analysis for OD. *Research in Organizational Change and Development*, Vol. 1, pp. 171-228.

Burgelman, R. (1983). A process model of internal corporate venturing in the diversified major firm. *Administrative Science Quarterly*, Vol. 30, pp. 223-244.

Burns, T. and Stalker, G.M. (1961). *The Management of Innovation*. London: Tavistock Press.

Burt, D.N. (1989). Managing suppliers up to speed. *Harvard Business Review*, July-August, pp. 127-135.

Business Week (1990). King Customer. *Business Week*, 12 March, pp. 88-94.

Business Week (1993). A wellspring of innovation. *Business Week*, 22 Oct., pp. 56-62.

Business Week (1994a). Unilever's global fight. *Business Week*, 4 July (3364-694), pp. 40-45.

Business Week (1994b). What's the word in the lab? Collaborate. *Business Week*, 27 June (3363-693), pp. 44-45.

Buzzel, R.D. and Gale, B.T. (1987). *The PIMS Principles: Linking Strategy to Performance*. New York: Free Press.

Calantone, R. and Cooper, R.G. (1979). A discriminant model for identifying scenarios of industrial new product failure. *Journal of the Academy of Marketing Science*, Vol. 7, pp. 163-183.

Calantone, R.J., Benedetto, C.A. di and Divine, R. (1993). Organisational, technical and market antecedents for successful new product development. *R&D Management*, Vol. 23 (4), pp. 337-351.

Caniëls, M. and Verspagen, B. (1992). R&D-intensiteit bij bedrijven: hoopvol of zorgwekkend? *ESB*, Vol. 77 (3880), pp. 978-979.

Cannell, W. and Dankbaar, B. (1996). *Technology Management and Public Policy in the European Union*. Oxford: Oxford University Press.

Centraal Bureau voor de Statistiek (1991). *Statistisch Jaarboek*. The Hague: CBS.

Centraal Bureau voor de Statistiek (1993). *Statistisch Bulletin*, 27, 8 July, p. 2.

Chandler, A. (1962). *Strategy and Structure: Chapters in the History of American Industrial Enterprise*. Cambridge, Mass.: Harvard University Press.

Chen, M.-J., Farh, J.-L. and MacMillan, I.C. (1993). An exploration of the expertness of outside informants. *Academy of Management Journal*, Vol. 36 (6), pp. 1614-1632.

Chew, W.B., Leonard-Barton, D. and Bohn, R.E. (1991). Beating Murphy's Law. *Sloan Management Review*, Vol. 32 (3), pp. 5-16.

Child, J. (1984). *Organisation: A guide to Problems and Practice*. London: Harper and Row.

Christopher, M., Payne, A. and Ballantyne, D. (1994). *Relationship Marketing: Bringing Quality, Customer Service, and Marketing Together*. Oxford: Butterworth-Heinemann.

Clare, D.A. and Sanford, D.G. (1984). Cooperation and conflict between industrial sales and production. *Industrial Marketing Management*, Vol. 13, pp. 163-169.

Clark, K.B. and Fujimoto, T. (1991). *Product Development Performance: Strategy, Organization, and Management in the World Auto Industry*. Boston, Mass.: Harvard Business School Press.

Clark, K.B. and Wheelright, S.C. (1993). *Managing New Product and Process Development*. New York: Free Press.

Cobbenhagen, J.W.C.M. (1997a) Cine/Vista: How an assessment of a new technology led to a strategic shift. In: D. Brown (ed.), *Innovation Management Tools: A Review of Selected methodologies*. Luxembourg: Office for Official Publications of the European Communities, pp. 312-318.

Cobbenhagen, J.W.C.M. (1997b) Product/market/technology scan. In: D. Brown (ed.), *Innovation Management Tools: A Review of Selected Methodologies*. Luxembourg: Office for Official Publications of the European Communities, pp. 137-146.

Cobbenhagen, J.W.C.M. (1990). Innoveren: strategieën en modellen. In J.F. den Hertog, F.M. van Eijnatten (eds), *Management van technologische vernieuwing*, Maastricht/Assen: van Gorcum, pp. 8-10.

Cobbenhagen, J.W.C.M., Dankbaar, B. and Wolters, A. (1996a). *De Vlaamse technologische infrastructuur vanuit de KMO-optiek bekeken*. Maastricht: UPM.

Cobbenhagen, J.W.C.M. and Hertog, J.F. den (1992). Technology assessment en bedrijfsbeleid. In R.T. Frambach and E. Nijssen (eds), *Technologie en Strategisch Management*. Culemburg: LEMMA.

Cobbenhagen, J.W.C.M. and Hertog, J.F. den (1995). Tomando la inciativa: lecciones de empresas innovadoras con exito en los Paises Bajos. *Economia Industrial*, Vol. 301, pp. 141-151.

Cobbenhagen, J.W.C.M., Hertog, J.F. den and Pennings, J.M. (1993). Innovating succesfully: breaking away from the pack. Paper presented at the Academy of Management 1993 Annual Meeting, Atlanta, August 6-11.

Cobbenhagen, J.W.C.M., Hertog, J.F. den and Pennings, J.M. (1994a). *Succesvol Veranderen: Kerncompetenties en Bedrijfsvernieuwing*. Deventer: Kluwer Bedrijfswetenschappen.

Cobbenhagen, J.W.C.M., Hertog, J.F. den and Pennings, J.M. (1994b). Sneller en beter innoveren voor minder geld: de ontwikkeling van kerncompetenties. *Holland Management Review*, Winter 94/95, pp. 99-112.

Cobbenhagen, J.W.C.M., Hertog, J.F. den and Pennings, J.M. (1995). Koplopers in bedrijfsvernieuwing. *Beleidsstudies Technologie Economie*, Vol. 29. The Hague: Ministerie van Economische Zaken.

Cobbenhagen, J.W.C.M., Hertog, J.F. den and Philips, G. (1990a). Management of innovation in the processing industry: a theoretical framework. In C. Freeman and L. Soete (eds), *New Explorations in the Economics of Technological Change.* London: Frances Pinter.

Cobbenhagen, J.W.C.M., Meeuwissen, R. and Sleuwaegen, L. (1996b). In entrepreneurial mode: Onderzoek naar innovatie in 400 Vlaamse ondernemingen. *Ondernemen,* Vol. 52 (4), pp. 6-12.

Cobbenhagen, J.W.C.M. and Philips, G. (1994). *Research and Technology Management in Enterprises: Issues for Community Policy (Sast project no.8). Case study in Chemicals.* Luxembourg: Office for Official Publications of the European Community.

Cobbenhagen, J.W.C.M., Philips, G. and Hertog, J.F. den (1990b*). Managing Innovations: Coping with Complexity and Dilemmas,* MERIT Research memorandum 90-010. Also paper presented at the Conference 'Firm Strategy and Technical Change: Microeconomics or Microsociology?' Manchester Business School/UMIST, 27-28 September.

Cobbenhagen, J. and Severijns.J. (1999). Towards a knowledge intensive regional economy: the RTP process in Limburg. In K. Morgan and C. Nauwelaers, *Regional Innovation Strategies,* London: Jessica Kingsley.

Cohen, M.D. (1991). Individual learning and organizational routine: emerging connections. *Organizational Science,* Vol. 2 (1), pp. 135-139.

Cohen, S.G. and Ledford, G.E. (1991). *The Effectiveness of Self Managing Teams: A Quasi Experiment.* CEO Publication G91-6(191*),* University of California, Los Angeles: Center for Creative Organizations.

Cohen, S.S. and Zysman J. (1988). Manufacturing innovation and American industrial competence. *Science,* Vol. 239, pp. 1110-1115.

Cohen, W.M. and Levinthal, D.A. (1989). Innovation and learning: the two faces and R&D. *Economic Journal,* Vol. 99.

Cohen, W.M. and Levinthal, D.A. (1990). Absorptive capacity: new perspectives on learning and innovation. *Administrative Science Quarterly,* Vol. 35 (2), pp. 128-152.

Collingridge, D. (1980). *The Social Control of Technology.* Londen: Frances Pinter.

Collis, D.J. (1991). A resource-based analysis of global competition: the case of the bearings industry. *Strategic Management Journal,* Vol. 12, pp. 49-68.

Colombo, U. (1980). A viewpoint on innovation and the chemical industry. *Research Policy,* Vol. 9, pp. 204-231.

Cook, T.D. and Campbell, D.T. (1979) *Quasi-experimentation: Design and Analysis Issues for Field Settings.* Boston, Mass.: Houghton Mifflin.

Cool, K. and Schendel, D. (1988). Performance differences among strategic group members. *Strategic Management Journal,* Vol. 9 (3), pp. 207-224.

Cooper, R.G. (1975). Why new industrial products fail. *Industrial Marketing Management,* Vol. 4, pp. 315-326.

Cooper, R.G. (1979). Identifying industrial new product success: Project NewProd. *Industrial Marketing Management,* Vol. 8, pp. 124-135.

Cooper, R.G. (1980). *Project NewProd: What Makes a New Product a Winner?* Montreal: Quebec Industrial Innovation Centre.

Cooper, R.G. (1983). The new product process: an empirically-based classification scheme. *R&D Management*, Vol. 13 (1), pp. 1-13.

Cooper, R.G. (1986). *Winning at New Products*. Reading, Mass.: Addison-Wesley.

Cooper, R.G. (1988). *Winning at New Products*. London: Kogan Page

Cooper, R.G. (1990). New products: what distinguishes the winners. *Research and Technology Management*, November-December, pp. 27-31.

Cooper, R.G. and Brentani, U. de (1984). Criteria for screening new industrial products. *Industrial Marketing Management*, Vol. 13, August, pp. 149-156.

Cooper, R.G. and Brentani, U. de (1991). New industrial financial services: what distinguishes the winners. *Journal of Product Innovation Management*, Vol. 8 (2), pp. 75-90.

Cooper, R. and Burrell, G. (1988). Modernism, post modernism and organizational analysis: an introduction. *Organization Studies*, Vol. 1.

Cooper, R.G. and Kleinschmidt, E.J. (1987a). New products: what separates winners from losers? *Journal of Product Innovation Management*, Vol. 4 (3), pp. 169-184.

Cooper, R.G. and Kleinschmidt, E.J. (1987b). What makes a new product a winner: success factors at the project level. *R&D Management*, Vol. 17 (3), pp. 175-189.

Cooper, R.G. and Kleinschmidt, E.J. (1993a). Major new products: what distinguishes the winners in the chemical industry? *Journal of Product Innovation Management*, Vol. 10 (2), pp. 90-111.

Cooper, R.G. and Kleinschmidt, E.J. (1993b). Screening new products for potential winners. *Long Range Planning*, Vol. 26 (6), pp. 74-81.

Cooper, R.G. and Kleinschmidt, E.J. (1994). Determinants of timeliness in product development. *Journal of Product Innovation Management*, Vol. 11, pp. 381-396.

Cooper, R.G., Easingwood, C.J., Edgett, S., Kleinschmidt, E.J. and Storey, C. (1994). What distinguishes the top performing new products in financial services. *Journal of Product Innovation Management*, Vol. 11, pp. 281-299.

Corey, E.R. and Starr, S.A. (1971). *Organization Strategy: A Marketing Approach*. Boston, Mass.: Harvard University Press.

Coriat, B. (1995). Organisational innovations: the missing link in European competitiveness. In L.E. Andreasen, B. Coriat, F. den Hertog and R. Kaplinsky, *Europe's Next Step*, London: Frank Cass, pp. 3-32.

Corvers, F., Dankbaar, B. and Hassink, R. (1994). Euregio's in Nederland. Een inventarisatie van economische ontwikkelingen en beleid. The Hague: Commissie Ontwikkeling Bedrijven/Sociaal Economische Raad.

Covalski, M.A. and Dirsmith, M.W. (1990). Dialectic tension, double reflexivity and the everyday accounting researcher: on using qualitative methods. *Accounting, Organizations and Society*, Vol. 15 (6), pp. 543-573.

Crawford, C.M. (1991). *New Products Management*. 3rd edn, Homewood, Ill.: Irwin.

Cresson, E. and Bengemann, M. (1995). *Green Paper on Innovation*. Brussels/ Luxembourg: ECSC-EC-EAEC.

Crosby, P.B. (1979). *Quality is Free*. New York: McGraw-Hill.

Cummings, W.T., Jackson, D.W. and Ostrom, L.L. (1984). Differences between industrial and consumer product managers. *Industrial Marketing Management*, Vol. 13, pp. 171-180.

Daey Ouwens, C., Hoogstraten, P. van, Jelsma, J., Prakke, F. and Rip, A. (1987). *Constructief Technologische Aspectenonderzoek, een verkenning*. The Hague: Staatsdrukkerij.

Daft, R.L. and Lewin, A.Y. (1993), Where are the theories for the 'new' organisational forms? An editorial essay. *Organization Science*, Vol. 4 (4), pp. i-vi.

Dalton, M. (1959). *Men Who Manage*. New York: John Wiley.

Dankbaar, B. (1993). *Research and Technology Management in Enterprises: Issues for Community Policy* (Sast project, No.8) Overall Strategic Review. Luxemburg: Commision of the European Communities.

Dankbaar, B. (1996). *Organiseren in een turbulente omgeving*. Oratie. Nijmegen: Nijmegen Business School.

Dankbaar, B. (1997). Organiseren in een turbulente omgeving. De creatieve ondernemening op de drempel van de 21ste eeuw. *Tijdschrift voor Politieke Ekonomie*, Vol. 19 (4), pp. 8-33.

Davis J.H. (1973). Group decision and social interaction: a theory of social decision schemes. *Psychological Reviews*, Vol. 80, pp. 97-125.

Davis, L.E. (1976). Developments in job design. In P. Warr (ed.), *Personal Goals and Work Design*, London: John Wiley.

Deal, T. and Kennedy, A. (1982). *Corporate Cultures: The Rites and Rituals of Corporate Life*. Reading, Mass.: John Wiley.

De Bono, E. (1971). *Lateral Thinking for Management*. England: American Management Association.

Dijkstra, S., Kerkhoff, W. and Simonse, A. (1986). Technology assessment. De ontwikkeling van nieuwe technologie op basis van sociale waarderingsgrondslag. *Tijdschrift voor Arbeidsvraagstukken*, Vol. 2 (3), pp. 45-55.

Docherty, P. (1992). The learning organization: Swedish concepts. Aims and experiences from two of the Swedish Work Environment Fund's R&D Programmes. In H. Pornschlegel, *Research and Development in Work and Technology*. Heidelberg: Physica-Verlag.

Docter, J., Horst, R. van der and Stokman, C. (1989). Innovation processes in small and medium-size companies. *Entrepreneurship and Regional Development*, Vol. 1, pp. 33-52.

Dodgson, M. (1993). Organizational learning: A review of some literatures. *Organization Studies*, Vol. 14 (3), pp. 375-394.

Donnellon, A. (1993). Crossfunctional teams in product development: accommodating the structure to the process. *Journal of Product Innovation Management,* Vol. 10 (5), pp. 377-392.

Doty, D.H. and Glick, W.H. (1994). Typologiea as a unique form of theory building: toward improved understanding and modelling. *Academy of Management Review*, Vol. 19 (2), pp. 230-251.

Dougherty, D. (1990). Understanding new markets for new products. *Strategic Management Journal*, Vol. 11-1, 59-78.

Dougherty, D. (1992). Interpretive barriers to successful product innovation in large firms. *Organisation Science*, Vol. 3 (2), pp. 179-202.

Dougherty, D. (1993). *Managing your Core Incompetencies for Product Innovation.* Working paper. Montreal: McGill University.

Dougherty, D.J. and Pennings, J.M. (1991). *Success Factors of Innovation.* Philadelphia: The Wharton School.

Dove, R. (1995). Agile practice reference models. *Production*, July.

DTI (Department of Trade and Industry) and CBI (Confederation of British Industry). (1993). *Innovation, the Best Practice.* London: Department of Trade and Industry.

Dunn, W.N. and Swierczek, F.W. (1977). Planned organizational change: toward grounded theory. *Journal of Applied Behavioral Science*, Vol. 13 (2), pp. 135-157.

Easingwood, C.J. and Storey, C. (1991). Success factors for new consumer financial services. *International Journal of Bank Marketing*, Vol. 9 (1), pp. 3-10.

Easterby-Smith, M. (1990). Creating a learning organisation. *Personnel Review*, Vol. 19 (5), pp. 24-28.

Eisenhardt, K.M. (1989). Building theories from case study research. *Academy of Management Review*, Vol. 14 (4), pp. 532-550.

Elfring, T. and Kloosterman, R.C. (1989). De Nederlandse 'Job machine'. *ESB*, Vol. 74, pp. 736-740.

Eliasson, G. (1990). The firm as a competent team. *Journal of Economic Behavior and Organization*, Vol. 13, pp. 275-298.

Enright, M.J. (1989). *Geographic Concentration, Vertical Structure and Innovation.* Working Paper, Cambridge, Mass.: Harvard University.

Ettlie, J.E. (1976). The timing and sources of information for the adoption and implementation of production innovation. *IEEE Transactions on Engineering Management*, Vol. 23, pp. 62-68.

Feldman, S.P. (1989). The broken wheel: the inseparability of autonomy and control within organisations. *Journal of Management Studies*, Vol. 26 (2), pp. 83-102.

FEM (1994). Valse start. *FEM*, 25 June, pp. 16-19.

Festinger, L. (1954). A theory of social comparison processes. *Human Relations*, Vol. 7, pp. 117-140.

Fletcher, R.H. and Fletcher, S.W. (1979). Clinical research in general medicine journals: a 30-year perspective. *New England Journal of Medicine,* Vol. 301, pp. 180-183.

Forbes, P.M. (1991). Marketing management: developing a core competence. *National Petroleum News*, Vol. 83 (5), p. 59.

Ford, D. and Ryan, C. (1981). Taking technology to market. *Harvard Business Review*. March-April, pp. 117-126.

Foster, R.N. (1986a). *Innovation: The Attacker's Advantage.* London: Pan Books.

Foster, R.N. (1986b). Timing technological transitions. In M. Horwitch (ed.) *Technology in the Modern Corporation*, New York, Pergamon, pp. 35-49.

Fowler, F.J. Jr (1988). *Survey Research Methods* rev. edn. Newbury Park, Cal.: Sage.

Franko, L.G. (1989). Global corporate competition: who's winning, who's losing, and the R&D factor as one reason why. *Strategic Management Journal*, Vol. 10, pp. 449-474.

Freeman, C. (1982). *The Economics of Industrial Innovation*, 2nd edn. London: Frances Pinter.

Freeman, C. (1988). De diffusie van informatietechnologie in de tertiaire sector. *TPE*, Vol. 11 (3), pp. 97-105.

Frohman, A.L. (1980). Managing the company's technological assets. *Research Management*, September, pp. 20-24.

Frontini, G.F. and Richardson, P.R. (1984). Design and demonstration: the key to industrial innovation. *Sloan Management Review*, Summer, pp. 29-39.

Gaito, J. (1980). Measurement scales and statistics: resurgence of an old misconception. *Psychological Bulletin*, Vol. 87, pp. 564-567.

Galbraith, J.R. (1971). Matrix organisation design: how to combine functional and project forms. *Business Horizons*, Vol. 17 (1), pp. 29-40.

Galbraith, J.R. (1984). Human resource policies for the innovating organisation. In C. Fombrun, N.M. Tichy and M.A. Devanna (eds), *Strategic Human Resource Management*, New York: John Wiley.

Galbraith, J.R. (1994). *Competing with Flexible Lateral Organisations*, 2nd edn. Reading, Mass.: Addison-Wesley.

Gammill, L.M. (1992). Organizational change: a case study. *Management Review*, Vol. 81 (11), p. 63.

Gannon, M.J., Smith, K.G. and Grimm, C. (1992). An organizational information-processing profile of first movers. *Journal of Business Research*, Vol. 25 (3), pp. 231-241.

Garratt, B. (1987). Learning is the core of organisational survival: action learning is the key integrating process. *Journal of Management Development*, Vol. 6 (2), pp. 38-44.

Geertz, C. (1973). *The Interpretation of Cultures*. New York: Basic Books.

Gehani, R.R. (1992). Concurrent product development for fast track corporations. *Long Range Planning*, Vol. 6 (25), pp. 40-47.

Gergen, K.J. (1990). Het binnenstebuiten. *Filosofie in bedrijf*, Vol. 2 (1), pp. 21-30.

Geroski, P. and Machin, S. (1992) Do innovating firms out-perform non-innovators? *Business Strategy Review*, Summer, pp. 79-90.

Gerpott, T.J. and Wittkemper, G. (1991). Verkürzung von Produktentwicklungs-zeiten: Vorgehensweise und Ansatzpunkte zum Erreichen Technologischer Sprintfähigkeit. In: Booz, Allen and Hamilton (ed.), *Integriertes Technologie- und Innovationsmanagement*, Berlin: E. Schmidt. pp. 117-145.

Gerstenfeld, A. and Sumiyoshi, K. (1980) The management of innovation in Japan. *Research Management*, Vol. 23, pp. 30-34.

Gerwin, D. and Guild, P. (1994). Redefining the new product introduction process. *International Journal of Technology Management*, Vol. 9 (5/6/7), pp. 678-690.

Ghobadian, A., Speller, S. and Jones, M. (1994). Service quality: concepts and models. *International Journal of Quality and Reliability Management*, Vol. 11 (9), pp. 43-66.

Glaser, B. and Strauss, A. (1967). *The Discovery of Grounded Theory: Strategies of Qualitative Research*. London: Weidenfeld and Nicolson.

Glick, W.H., Huber, G.P., Miller, C.Ch., Doty, D.H. and Sutcliff, K.M. (1990). Studying changes in organizational design and effectiveness: retrospective event histories and periodic assessment. *Organization Science*, Vol. 1 (3), pp. 293-312.

Gobeli, D.H. and Brown, D.J. (1987). Analyzing product innovations. *Research Management*, July-August, pp. 25-31.

Gobeli, D.H. and Larson, E.W. (1986) Matrix management: more than a fad. *Engineering Management International.* Vol. 4, pp. 71-76.

Golder, P.N. and Tellis, G.J. (1993) Pioneer advantage: marketing logic or marketing legend? *Journal of Marketing Research*, Vol. 30 (2), pp. 158-170.

Goldman, S.L., Roger, N.N. and Preiss, K. (1995). *Agile Competitors and Virtual Organizations.* New York: Van Nostrand Reinhold.

Gorter, T. (1994). Kerncompetenties: de brug tussen huidige business en toekomstige identiteit. In A. Witteveen (ed.), *Top 20 Trends in Strategisch Management.* Amsterdam: Management Press, pp. 124-130.

Graaff, L.E. de and Jaspers, P.F.M. (1989). *Overwinnen bij brancheproblemen.* The Hague: Nationale Investeringsbank.

Grant, R.M. (1991). The resource-based theory of competitive advantage: implications for strategy formulation. *California Management Review*, Spring, pp. 114-135.

Griliches, Z. (1958). Hybrid corn: an exploration of the economics of technological change. *Econometrica*, Vol. 25, pp. 501-522.

Grindley, P. (1989). Technological change within the firm: a framework for research and management. Paper presented at the third annual conference of the British Academy of Management, Manchester, 10-12 Sept.

Grønhaug, K. and Kaufmann, G. (1988). *Innovation: A Cross-disciplinary Perspective.* Oslo: Norwegian University Press.

Grönlund, A. and Jönsson, S. (1989). Managing for cost improvement in automated production. Paper presented at the colloquium on 'Measuring Manufacturing Performance', Harvard University Graduate School of Business Administration, 25-26 January.

Grunow, D. (1995). The research design in organization studies: problems and prospects. *Organization Science*, Vol. 6 (1), pp. 93-103.

Gupta, A.K. and Wilemon, D. (1988). The credibility-cooperation connection at the R&D-marketing interface. *Journal of Product Innovation Management,* Vol. 5, pp. 20-31.

Hage, J. (1980). *Theories of Organizations: Form, Process, and Transformation.* New York and Chichester, Sussex: John Wiley.

Hambrick, D.C. (1983). An empirical typology of mature industrial product environment. *Academy of Management Journal*, Vol. 26, pp. 213-230.

Hamel, G. and Prahalad, C.K. (1989). Strategic intent. *Harvard Business Review*, May-June, pp. 63-76.

Hamel, G. and Prahalad, C.K. (1991). Corporate imagination and expeditionary marketing. *Harvard Business Review*, July-August, pp. 81-92.

Hamel, G. and Prahalad, C.K. (1994a). Competing for the future. *Harvard Business Review*, July-August, pp. 122-128.

Hamel, G. and Prahalad, C.K. (1994b). *Competing for the Future*, Boston, Mass.: HBS Press.

Hammersley, M. and Atkinson, P. (1983). *Ethnography: Principles in Practice.* London: Tavistock Press.

Hannan, M.T. and Freeman, J.H. (1990). *Organizational Ecology*, New York: Harper.

Hansen, G. and Wernerfelt, B. (1989). Determinants of firm performance: the relative importance of economic and organisational factors. *Strategic Management Journal*, Vol. 10, pp. 399-411.

Harrigan, K.R. (1983). Research methodologies for contingency approaches to business strategy. *Academy of Management Review*, Vol. 8, pp. 398-405.

Hartgers, H., Dagevos, J., Oerlemans, L. and Boekema, F. (1990). *Responsie, een Literatuurstudie naar Nieuwe Aangrijpingspunten voor Onderzoek naar Regionale-ekonomische Specialisatie en Ontspecialisatie*. Tilburg: Economisch Instituut Tilburg.

Hays, W.L. (1988). *Statistics*. Forth Worth, Tx.: Rinehart and Winston.

Hedlund, G. (1994). A model of knowledge management and the N-form corporation. *Strategic Management Journal*, Vol. 15, pp. 73-90.

Helleloid, D. and Simonin, B. (1992). Organizational learning and a firm's core competence. Paper presented at the workshop 'Competence-based Competition: Towards the Roots of Sustainable Competitive Advantage, Genk.

Henderson, R. and Cockburn, J. (1994). Measuring competence? Exploring firm effects in pharmaceutical research. *Strategic Management Journal*, Vol. 15, pp. 63-84.

Henderson, R.M. and Clark, K.B. (1990). Architectural innovation: The reconfiguration of existing product technologies and the failure of established firms. *Administrative Science Quarterly*, Vol. 35, pp. 9-30.

Henke, J.W., Krachenberg, A.R. and Lyons, T.F. (1993). Perspective. cross-functional teams: good concept, poor implementation! *Journal of Product Innovation Management* , Vol. 10, pp. 216-229.

Hershey, J.C., Kunreuther, H.C. and Schoemaker, P.J. (1988). Sources of bias in assessment procedures for utility functions. In: D.A. Bell, H. Raiffa and A. Tversky (eds), *Decision Making: Descriptive, Normative, and Prescriptive Interactions*. Cambridge: Cambridge University Press, pp. 422-442.

Hertog, J.F. den (1979). *Arbeitsstrukturierung: Experimente aus Holland*. Bern: Huber.

Hertog, J.F. den (1991). *Technology, Work and Organisation: a joint link between FAST and MODEM on Anthropocentric Technologies. Design, Development and Diffusion*. Report for the European Commision. MONITOR-FAST work programme 1989-1990.

Hertog, J.F. den (1992). Policy considerations for anthropocentric research and development. In H. Pornschlegel (ed.), *Research and Development in Work and Technology*. Heidelberg: Physica-Verlag, pp. 167-192.

Hertog, J.F. den, and Cobbenhagen, J.W.C.M. (1995). Laat je niet op achterstand fietsen (Heeft de Nederlandse R&D nog goede benen?). *PolyTechnisch Tijdschrift*, Vol. 50 (10), pp. 22-24.

Hertog, J.F. den, and Diepen, S.J.B. van (1989). Technological innovation and organisational learning. In H.J. Bullinger (ed.), *Information Technology for Organizational Systems*. Amsterdam: North-Holland.

Hertog, J.F., den, and Eijnatten, F.M. van (eds) (1990*). Management van technologische vernieuwing*. Maastricht/Assen: van Gorcum.

Hertog, J.F. den and Huizenga, E. (1997). *De kennisfactor*. Deventer: Kluwer bedrijfsinformatie.

Hertog, J.F. den and Roberts, H.J.E. (1992). Learning strategies for management accounting in unprogrammable contexts. *Accounting, Management and Information Technology*, Vol. 2 (3), pp. 165-182.

Hertog, J.F. den and Schröder, P. (1989). *Social Research for Technological Change: Lessons from National Programmes in Europe*. MERIT RM 89-028, Maastricht.

Hertog, J.F. den, and Sluijs, E. van (1995). *Onderzoek in Organisaties. Een methodologische reisgids*. Assen: van Gorcum.

Hertog, J.F. den and Wielinga, C. (1992). Control systems in dissonance: the computer as an inkblot. *Accounting, Organizations and Society*, Vol. 17 (2), 103-127.

Hertog, J.F. den, Cobbenhagen, J.W.C.M. and Philips, G. (1990). *Thriving Innovation Beyond the Myths of Functional Concentration and Organisational Consonance*. Maastricht: MERIT Research memorandum 90-013. Also paper presented at the International Conference on Technology Transfer and Innovation in Mixed Economies, TTI '90, 27-29 August, Trondheim, Norway.

Hertog, J.F. den, Sluijs, E. van, Diepen, B. van, Assen, A. van (1991). Innovatie en personeelsbeleid: de beheersing van de kennishuishouding. *Bedrijfskunde*, Vol. 63, (2).

Hertog, J.F. den, Philips, G. and Cobbenhagen, J.W.C.M. (1996). Paradox Management: the fourth phase in innovation management. In A. Gutschelhofer and J. Scheff (eds), *Paradoxes Management. Widersprüche im management, management der widersprüche*. Vienna: Linde Verlag, pp. 43-77.

Hirsh, R.F. (1986). How success short-circuits the future. *Harvard Business Review*, March-April, pp. 72-76.

Hise, R.T., O'Neal, L., Parsuraman, A. and McNeal, J.U. (1990). R&D interaction in new product development: implications for new product success rates. *Journal of Product Innovtion Management*, Vol. 7, pp. 142-155.

Hoesel, P.H.M. van (1985). *Het Programmeren van Sociaal Beleidsonderzoek: Analyse en Receptuur*. Leiden: LISBON.

Hofer, C.W. and Schendel, D. (1978). *Strategy Formulation: Analytical Concepts*. St Paul, Minn.: West Publishing.

Hoffman, L.R. (ed.) (1979). *The Group Problem Solving Process: Studies of a Valence Model*. New York: Praeger.

Hofstede, G., Bond, M. and Luk, C. (1993). Individual perceptions of organizational cultures: a methodological treatise on levels of analysis. *Organization Studies*, Vol. 14 (4), pp. 483-503.

Homans, G.C. (1950). *The Human Group*. New York: Harcourt Brace.

Hopkins, D.S. and Bailey, E.L. (1971). New product pressures. *Conference Board Records*, Vol. 8, pp. 16-24.

Huber, G.P. (1991). Organizational learning: the contributing processes and the literatures. *Organization Science*, Vol. 2 (1), pp. 88-115.

Hutt, M.D. and Speh, T.W. (1984). The marketing strategy center: diagnosing the industrial marketer's interdisciplinary role. *Journal of Marketing*, Vol. 48 (4), pp. 53-61.

Huxley, A. (1963). Achieving a perspective on the technological order. *Technology and Culture*, Vol. 3, pp. 636-642.

Imai, K., Nonaka, I. and Takeuchi, H. (1985). Managing the new product development process: how Japanese companies learn and unlearn. In K. Clark, R. Hayes and C. Lorenz (eds), *The Uneasy Alliance*, Boston, Mass.: Harvard Business School Press.

Innovatiecentra, KPMG and NMB Bank (1991). *Samen Innoveren. Technologische samenwerking bij het midden- en kleinbedrijf: de economische voorwaarde.* Haarlem: Innovatiecentrum.

Irvin, R.A. and Michaels, E. (1989). Core skills: doing the right things right. *McKinsey Quarterly*, Summer, pp. 4-19.

Itami, H. (1987). *Mobilizing Invisible Assets.* Cambridge, Mass.: Harvard University Press.

Jacobson, R. (1988). The persistence of abnormal returns. *Strategic Management Journal*, Vol. 9 (1), pp. 41-58.

Jahoda, M., Deutsch, M. and Cook, S.W. (1951). *Research Methods in Social Relations.* New York: Dryden Press.

Jaikumar, R. (1986). Postindustrial manufacturing. *Harvard Business Review*, Vol. 64 (6), pp. 69-76.

Jelinek, M. and Bird Schoonhoven, C. (1990). *The Innovation Marathon.* Oxford: Basil Blackwell.

Johne, A. and Snelson, P. (1988). Success factors in product innovation: a selected review of the literature. *Journal of Product Innovation Managment*, Vol. 5 (2), pp. 100-110.

Johnson, G. and Scholes, K. (1993). *Exploring Corporate Strategy*, London: Prentice-Hall.

Jong, W.M. de (1991). *Perspectief in Innovatie, de Chemie nader beschouwd.* The Hague: SDU, WRR T-1.

Jovanovic, B. and MacDonald, G.M. (1994). Competitive diffusion. *Journal of Political Economy,* Vol. 102 (1), pp. 24-52.

Juran, J.M. (1974). *Quality Control Handbook.* New York: McGraw-Hill.

Kaiser, H.F. (1974). An index of factorial simplicity. *Psychometrika.* Vol. 39, pp. 31-36.

Kamien, M. and Schwartz, N. (1982). *Market Structure and innovation.* Cambridge: Cambridge University Press.

Kandel, N., Remy J.P., Stein, C. and Durand, T. (1991). Who's who in technology: identifying technological competence within the firm. *R&D Management*, Vol. 21 (3), pp. 215-228.

Kantrow, A.M. (1980). The strategy-technology connection. *Harvard Business Review.* July-August, pp. 6-21.

Kaplan, R.S. (1986a). Must CIM be justified by faith alone? *Harvard Business Review,* Vol. 2, pp. 87-95.

Kaplan, R.S. (1986b). The role of empirical research in management accounting. *Accounting, Organizations and Society*, Vol. 11 (4/5), pp. 429-452.

Karakaya, F. and Kobu, B. (1993). New product development process: an investigation of success and failure in high-technology and non-high-technology firms. *Journal of Business Venturing*, Vol. 9, pp. 49-66.

Katz, D. and Kahn, R.L. (1978). *The Social Psychology of Organizations*. New York: John Wiley.

Katz, R. and Allen, T. (1985). Project performance and the locus of influence in the R&D matrix. *Academy of Management Journal*, Vol. 28 (1), pp. 67-87.

Keller, R.T. (1986). Predictors of the performance of project groups in R&D organisations. *Academy of Management Journal*, Vol. 29 (4), pp. 715-726.

Kerin, R.A., Varadarajan, P.R. and Peterson, R.A. (1992). First-mover advantage: a synthesis, conceptual framework, and research propositions. *Journal of Marketing*, Vol. 56 (4), pp. 33-52.

Kidder, L.H. and Judd, Ch.M. (1986). *Research Methods in Social Relations,* 5th edn. New York: CBS Publishing Japan.

Kilmann, R., Saxton, M., Serpa, R. and Associates (1985*). Gaining Control of the Corporate Culture*, San Francisco: Jossey-Bass.

Kim, D.H. (1993). The link between individual and organizational learning. *Sloan Management Review*, Fall, pp. 37-50.

Kleinknecht, A. (1990). *Diffusie van Technologische Vernieuwing*, The Hague: Stichting voor Economisch Onderzoek.

Klimstra, P.D. and Potts, J. (1988). What we've learned about managing R&D projects. *Research and Technology Management*, May-June, pp. 42-58.

Kline, S.J. (1985). Innovation is not a linear process. *Research Management*, July-August, pp. 36-45.

Kogut, B. (1988). Joint ventures: theoretical and empirical perspectives. *Strategic Management Journal*, Vol. 9 (4), pp. 319-332.

Kogut, B. and Zander, U. (1992). Knowledge of the firm, combinative capabilities, and the replication of technology. *Organization Science*, Vol. 3 (3), pp. 383-397.

Krachenberg, A.R., Henke, J.W. Jr and Lyons, T.F. (1988). An organisational structure response to competition. In G.E. Lasker (ed.), *Advances in Systems Research and Cybernetics,* Ontario: University of Windsor, pp. 320-326.

Krippendorff, K. von (1989). *Content Analysis: an Introduction to its Methodology.* Beverly Hills, Cal.: Sage.

Krubasik, E.G. (1988). Customize your product development. *Harvard Business Review*, November-December.

Kusnier, K. (1992). Marketing research: using tradition to move forward. *LIMRA's Market Facts*, Vol. 11 (2), pp. 30-33.

Lakatos, I. (1970). Falsification and the methodology of scientific research programmes. In I. Lakatos and M. Musgrave (eds), *Critisism and growth of Knowledge*, New York: Cambridge University Press.

Lambkin, M. (1988). Order of entry and performance in new markets. *Strategic Management Journal*, Vol. 9 (Special Issue), pp. 127-140.

Lammers, C.J. (1983). *Organisaties Vergelijkenderwijs*. Utrecht: Het Spectrum.

Lammers, C.J. (1986). De excellente onderneming als organisatiemodel. *Harvard Holland Review*, Autumn, 8, pp. 18-27.

Lansley, P.R. and Hillebrandt, P.M. (1996). Managing the construction enterprise: the need for better theory. Paper presented at the CIB W89 Beijing International Conference, 21-24 October.

Lant, T.K. and Mezias, S.J. (1990). Managing discontinuous change: a simulation study of organizational learning and entrepreneurship. *Strategic Management Journal*, Vol. 11 (Special Summer Issue), pp. 147-179.

Larson, E.W. and Gobeli, D.H. (1985). Project management structures: is there a common language? *Project Management Journal*, Vol. 16 (2), pp. 40-44.

Larson, E.W. and Gobeli, D.H. (1988). Organizing for product development projects. *Journal of Product Innovation Management*, Vol. 5, pp. 180-190.

Larsson, R. (1993). Case survey methodology: qualitative analysis of patterns accross case studies. *Academy of Management Journal*, Vol. 36 (6), pp. 1515-1546.

Lawrence, P.R. and Lorsch, J.W. (1967). *Organization and Environment: Managing Differentiation and Integration*. Homewood, Ill.: R.D. Irwin.

Lawrence, P.R. and Lorsch. T. (1969). *Developing Organisations: Diagnosis and Actions*. Reading, Mass.: Addison-Wesley.

Leest, D.J.B. and Weel, A.J. van der (1990). *Rapport van een pilot projekt op het gebied van Human Resource Management bij de business eenheid speciale producten*. 'Chemco' internal report.

Lemaitre, N. and Stenier, B. (1988). Stimulating innovation in large companies: observations and recommendations for Belgium. *R&D Management*, Vol. 18 (2), pp. 141-157.

Lemmink, J. (1991). *Kwaliteitsconcurrentie tussen ondernemingen*. PhD thesis. Maastricht: Rijksuniversiteit Limburg.

Leonard-Barton, D. (1990). A dual methodology for case studies: synergistic use of a longitudinal single site with replicated multiple sites. *Organization Science*, Vol. 1 (3), pp. 248-266.

Leonard-Barton, D. (1991). The role of process innovation and adaptation in attaining strategic technological capability. *International Journal of Technology*, Vol. 6 (3-4).

Leonard-Barton, D. (1992). Core capabilities and core rigidities: a paradox in managing new product development. *Strategic Management Journal*, Vol, 13, pp. 111-125.

Leonard-Barton, D. (1995). *Wellsprings of Knowledge*. Boston, Mass.: Harvard Business Press.

Leonard-Barton, D. and Kraus, W.A. (1985). Implementing new technology. *Harvard Business Review*, November-December, pp. 102-110.

Leplat, J. (1990). Skills and tacit skills: a psychological perspective. *Applied Psychology*, Vol. 39 (2), pp. 143-154.

Levinthal, D. (1992). Wisdom from Wharton: R&D as investment in learning. *Chief Executive*, no. 81 (Nov./Dec.), pp. 62-64.

Lewandowski, J. and O'Toole, J. (1990). *Forming the future: the marriage of people and technology at Saturn*. Stanford University, March, 29.

Lewis, J.D. (1991). Competitive alliances redefine companies. *Management Review*, Vol. 80 (4), pp. 14-18.

Lieberman, M.B. and Montgomery, D.B. (1988). First-mover advantages. *Strategic Management Journal*, Vol. 9, pp. 41-58.

Liemt, G.E. van and Commandeur, H.R. (1993). Duel in strategie: Porter versus Prahalad en Hamel. *Holland Harvard Review*, Vol. 36, pp. 113-119.

Likert, R. (1932). *A Technique for the Measurement of Attitudes*. Archives of Psychology, 140, Columbia University Press, pp. 44-53.

Likert, R. (1967). *The Human Organization: Its Management and Value*. New York: McGraw-Hill.

Link, P. (1987). Key to new product success and failure. *Industrial Marketing Management,* Vol. 16 (2), pp. 109-118.

Lomans, P. (1994). Troetelkinderen van de genetica. *Intermediair*, Vol. 30 (26), pp. 36-37.

Lounamaa, P.H. and March, J.G. (1987). Adaptive coordination of a learning team. *Management Science*, Vol. 33 (1), pp. 107-123.

Lundberg, C.C. (1988). Working with culture. *Journal of Organizational Change Management*. Vol.1 (2), pp. 38-47.

Lundvall, B.A. (ed.) (1992). *National Systems of Innovation: Towards a Theory of Innovation and Interactive Learning*. London: Frances Pinter.

Lyles, M.A. (1988). Learning among joint venture sophisticated firms. *Management International Review*, Vol. 28 (Special Issue), pp. 85-98.

Maanen, J. van and Barley, S.R. (1984). Occupational communities: culture and control in organisations. In B.M. Staw and L.L. Cummings (eds), *Research in Organizational Behavior*, Vol. 6, Greenwich, Conn.: JAI Press.

Mabert, V.A., Muth, J.F. and Schmenner, R.W. (1992). Collapsing new product development times: six case studies. *Journal of Product Innovation Management*, Vol. 9 (3), pp. 200-212.

Maidique, M.A. and Hayes, R.H. (1984). The art of high-technology management. *Sloan Management Review*, Winter, pp. 17-31.

Maidique, M.A. and Zirger, B.J. (1984). A study of success and failure in product innovation: The case of the US electronics industry. *IEEE Trans. Engineering Management*, Vol. EM-31, pp. 192-203.

Maidique, M.A. and Zirger, B.J. (1985). The new product learning cycle. *Research Policy*, Vol. 14, pp. 299-313.

Mann, F.C. (1957). Studying and creating change: a means to understanding social organization. Research in industrial human relations. *Industrial Relations Association*, Vol. 17, pp. 146-167.

Mansfield, D. (1989). The speed and cost of industrial innovation in Japan and the United States: external versus internal technology. *Management Science*, 34, pp. 1157-1173.

Mansfield, E. (1968). *Industrial Research and Technological Innovation*. New York: W.W. Norton.

Mantel, S.J. and Meredith, J.R. (1986). The role of customer cooperation in the development, marketing, and implementation of innovations. In H. Hübner (ed.), *The Art and Science of Innovation Management*, Amsterdam: Elsevier Science Publishers, pp. 27-36.

March, J.G., Sproul, L.S. and Tamuz, M. (1991). Learning from samples of one or fewer. *Organization Science*, Vol. 2, pp. 1-13.

Marquis, G. and Straight, D.M. (1965). *Organizational Factors in Project Performance*. Washington, D.C.: NASA.

Marsh, C. (1982). *The Survey Method: The Contribution of Surveys to Sociological explanation*. London: Allen and Unwin.

Mathôt, G.B.M. (1982). How to get new products to market quicker. *Long Range Planning*, Vol. 15 (6), pp. 20-30.

McCutcheon, D.M., Raturi, A.S. and Meredith, J.R. (1994). The customization-responsiveness squeeze. *Sloan Management Review*, Winter, pp. 89-99.

McDonough, E.F. and Barczak, G. (1991). Speeding up new product development: the effects of leadership style and source of technology. *Journal of Product Innovation Management*, Vol. 8 (3), pp. 203-211.

McGill, M.E. and Slocum, J.W. (1993). Unlearning the organization. *Organizational Dynamics*, Vol. 22 (2), pp. 67-79.

McGill, M.E., Slocum, J.W. and Lei, D. (1992). Management practices in learning organizations. *Organizational Dynamics*, Vol. 21 (1), pp. 5-17.

McGrath, J.E., Martin, J. and Kulka, R.A. (1982) *Judgement Calls in research*. Newbury Park, Cal.: Sage.

McKee, D. (1992). An organizational learning approach to product innovation. *Journal of Product Innovation Management*, Vol. 9 (3), pp. 232-245.

Meyer, A. de (1983). Management of technology in traditional industries: a pilot study in ten Belgian companies. *R&D Management*, Vol. 13 (1), pp. 15-22.

Meyer, A. de (1985). The flow of technological innovation in an R&D department. *Research Policy*, Vol. 14, pp. 315-328.

Meyer, A.D., Tsui, A.S. and Hinings, C.R. (1993). Configurational approaches to organizational analysis. *Academy of Management Journal*, Vol. 36 (6), pp. 1175-1195.

Meyer, M.H. and Roberts, E.B. (1988). Focusing product technology for corporate growth. *Sloan Management Review*, Summer, pp. 7-16.

Meyers, P.W. (1990). Non-linear learning in large technological firms: period four implies chaos. *Research Policy*, Vol. 19 (2), pp. 97-115.

Michell, J. (1986). Measurement scales and statistics: a clash of paradigms. *Psychological Bulletin*, Vol. 100, pp. 398-397.

Ministry of Economic Affairs (1994). *Indicators, Science and Technology*. The Hague: Dutch Ministry of Economic Affairs.

Mintzberg, H. (1979). *The Structuring of Organizations*. Englewood Cliffs, N.J.: Prentice-Hall.

Mintzberg, H. (1983). *Structures in Fives: Designing Effective Organizations*. Englewood Cliffs, N.J.: Prentice-Hall.

Mohr, L.B. (1982). *Explaining Organizational Behaviour: The Limits and Possibilities of Theory and Research*. San Francisco: Jossey Bass.

Moore, D.H. (1973). Evaluation of five discriminant procedures for binary variables. *Journal of the American Statistical Association*, Vol. 68, p. 399.

Morbey, G.K. (1988). R&D: its relationship to company performance. *Journal of Product Innovation Management*, Vol. 5, pp. 191-200.

Morgan, G. (1986). *Images of Organization*. Beverly Hills, Cal.: Sage.

Morgan, G. (1992). Improving your flysight. *Business Quarterly*, Vol. 57 (2), pp. 39-40.

Moss, S. (1981). *The Economic Theory of Business Strategy*. Sydney: Halstead Press.

Moss Kanter, R.M. (1983). *The Change Masters: Innovation for Productivity in the American Corporation*. New York: Simon and Schuster.

Moss Kanter, R. (1988). When a thousand flowers bloom: structural, collective and social conditions for innovation in organizations. In B. Staw and L. Cummings (eds), *Research in Organizational Behavior*. Greenwich, Conn.: JAI Press.

Moss Kanter, R. (1989a). *When Giants Learn to Dance*. New York: Simon and Schuster.

Moss Kanter, R. (1989b). Swimming in newstreams: mastering innovation dilemmas. *California Management Review*, Vol. 31 (4), pp. 45-69.

Moss Kanter, R. (1989c). Becoming PALs: pooling, allying and linking across companies. *Academy of Management Executive*, Vol. III (3), pp. 183-193.

Mulder, M. (1990). *Stimulerend leiderschap... is te stimuleren*. Nationaal Leiderschapscongres, 4 October, Rotterdam.

Mumford, E. and MacDonald, W.B. (1989). *XSEL's Progress: The Continuing Journey of an Expert System*. Chichester, Sussex: John Wiley.

Murman, P.A. (1994). Expected development time reductions in the German mechanical engineering industry. *Journal of Product Innovation Management*, Vol. 11 (3), pp. 236-252.

Myers, S. and Marquis, D.G. (1969). *Successful Industrial Innovations*. Washington, D.C.: National Science Foundation.

Nachmias, D. and Nachmias, C. (1976). *Research Methods in the Social Sciences*. New York: St. Martin's Press.

Nelson, R.R. and Winter, S.G. (1982). *An Evolutionary Theory of Economic Change*. Cambridge, Mass.: Harvard University Press.

NIPG (1958). *Hoe denkt U over uw werk?* Leiden: Nederlands Instituut voor Preventieve Geneeskunde.

Nonaka, I. (1988). Creating organizational order out of chaos: self renewal in Japanese firms. *California Management Review*, Vol. 30 (3), pp. 57-73.

Nonaka, I. (1990). redundant, overlapping organization: a Japanese approach to managing the innovation process. *California Management Review*, Vol. 32 (3), pp. 27-38.

Nonaka, I. (1991). The knowledge-creating company. *Harvard Business Review*, November-December, pp. 96-104.

Nonaka, I. (1994). A dynamic theory of organizational knowledge creation. *Organization Science*, Vol. 5 (1), pp. 14-37.

Nordhaug, O. and Grønhaug, K. (1994). Competences as resources in firms. *International Journal of Human Resource Management*, Vol. 5 (1), pp. 89-105.

Nunnally, J.C. and Bernstein, J.C. (1994). *Psychometric Theory*, 3rd edn. New York: McGraw-Hill.

OECD (1978). *Technology on Trial*. Paris: OECD.

O'Hare, M. (1988). *Innovate!* Oxford: Basil Blackwell.

Ohmae, K. (1982). *The Mind of the Strategist*. Harmondsworth, Middx.: Penguin.

Oijen, P.M.M. van (1987). *Beleid voor kennis, Een bijdrage tot een methodologie van de vraag*. Utrecht: Van Arkel.

Olsen, P. (1975). *Equipment Supplier-Producer Relationships and Process Innovations in the Textile Industry*. Boston, Mass.: Harvard Business Graduate School.

Olshavsky, R. (1980). Time and the rate of adoption of innovations. *Journal of Consumer Research*, Vol. 6, pp. 425-428.

Osborn, A.F. (1957). *Applied Imagination* rev.edn. New York: Scribner's.

Ott, J. (1989). *The Organizational Culture Perspective*. Pacific Grove, Cal.: Cole.

PA Consulting Group (1991). *Organising Product Design and Development*. London: Department of Trade and Industry/Design Council.

Page, A.L. (1993). Assessing new product development practices and performance: establishing crucial norms. *Journal of Product Innovation Management*, Vol. 10 (4), pp. 273-290.

Pavitt, K. (1984). Sectoral patterns of technical change: towards a taxonomy and a theory. *Research Policy*, Vol. 13, pp. 343-373.

Pavitt, K. (1987). Commentary on Tushman and Anderson's technological discontinuities and organization environments. In A.M. Pettigrew (ed.), *The Management of Strategic Change*, Oxford: Basil Blackwell, 1987, pp. 123-127.

Pavitt, K., Robson, M. and Townsend, J. (1987). The size distribution of innovating firms in the UK: 1945-1983. *Journal of Industrial Economics*, Vol. 45, pp. 297-306.

Payne, S.L. (1951). *The Art of Asking Questions*. Princeton, N.J.: Princeton University Press.

Pennings, J.M. and Harianto, F. (1992). The diffusion of technological innovation in the commercial banking industry. *Strategic Management Journal*, Vol. 13 (1), pp. 29-46.

Pennings, J.M., Cobbenhagen, J.W.C.M. and Hertog, J.F. den (1991). *Innoverende ondernemingen: Koplopers en achterblijvers*, Position paper. Maastricht: MERIT.

Penrose, E.T. (1959). *The Theory of the Growth of the Firm*, Oxford: Basil Blackwell.

Perry, L.T. (1990). Two virtues of competition. *Executive Excellence*, Vol. 7 (8), pp. 15-16.

Perry, T.S. (1990). Teamwork plus technology cuts development times. *IEEE Spectrum*, Vol. 27 (10), pp. 61-67.

Peteraf, M. (1993). The cornerstones of competitive advantage: a resource based view. *Strategic Management Journal*, Vol. 14 (3), pp. 179-191.

Peters, T. (1988a). *Thriving on Chaos: Handbook for a Management Revolution*. London: Macmillan.

Peters, T. (1988b). The mythodology of innovation, or a skunkworks tale, Part II. In M.L. Tushman and W.L. Moore (eds), *Readings in the Management of Innovation*, Cambridge, Mass.: Ballinger, pp. 138-147.

Peters, T. (1991). Get innovative or get dead, Part 1. *Engineering Management Review*, 19 (4), pp. 4-11.

Peters, T. (1992). Get innovative or get dead, Part 2. *Engineering Management Review*, 19 (5), pp. 7-14.

Peters, T.H. and Waterman, R.H. (1982). *In Search of Excellence: Lessons from America's Best Run Companies*. New York: Harper and Row.

Pettigrew, A. (1988). *The Management of Strategic Change*. Oxford: Basil Blackwell.

Philips, G., Hertog, J.F. den, Cobbenhagen, J. (1994) *Paradox Management: Product Development in the Processing Industry*. Maastricht: MERIT RM 2/94-011.

Pine, B.J. and Pietrocini, T.W. (1993). Standard modules allow mass customization at Bally Engineered Structures. *Planning Review*, Vol. 21 (4), pp. 20-22.

Pinto, M.B. (1989). Predictors of cross-functional cooperation in the implementation of marketing decisions. *AMA Educator's Proceedings*, Vol. 55, pp. 154-158.

Pinto, M.B. and Pinto, J.K. (1990). Project team communication and cross-functional cooperation in new program development. *Journal of Product Innovation Management*, Vol. 7, pp. 200-212.

Piore, M.J. and Sabel, C.F. (1984). *The Second Industrial Divide*. New York: Basic Books.

Porter, M.E. (1980). *Competitive Strategy: Techniques for Analyzing Industries and Competitors*. New York: Free Press.

Porter, M.E. (1985). *Competitive Advantage: Creating and Sustaining Superior Performance*. New York: Free Press.

Porter, M.E. (1988). From competitive advantage to corporate strategy. *Harvard Business Review*, Vol. 65. (May-June), pp. 43-59.

Porter, M.E. (1990). *The Competitive Advantage of Nations*. New York: Free Press.

Prahalad, C.K. (1990). The changing nature of worldwide competition: reversing the United States' decline. *Vital Speeches*, Vol. 56 (12), pp. 354-357.

Prahalad, C.K. (1993). The role of core competence of the corporation. *Research Technology Management*, Vol. 36, pp. 40-47.

Prahalad, C.K. and Hamel, G. (1990). The core competence of the corporation. *Harvard Business Review*, Vol. 68 (3), pp. 79-91.

Prigogine, I. and Stengers, I. (1984). *Order Out of Chaos: Man's New Dialogue with Nature*. London: Heinemann.

Pucik, V. (1988). Strategic alliances, organizational learning, and competitive advantage: The HRM Agenda. *Human Resource Management*, Vol. 27 (1), pp. 77-93.

Pugh, D.S., Hickson, D.J., Hinings, C.R. and Turner, C. (1968). Dimensions of organization structure. *Administrative Science Quarterly*, Vol. 13, pp. 65-105.

Qualls, W., Olshavsky, R. and Michaels, R. (1981). Shortening the PLC: an empirical test. *Journal of Marketing*, Vol. 45, pp. 76-80.

Quinn, J.B. (1985). Managing innovation: controlled chaos. *Harvard Business Review*, May-June, pp. 73-84.

Quinn, J.B. (1988a). General Motors Corporation: the downsizing decision. In J.B. Quinn, H. Mintzberg and R.M. James, *The Strategy Process: Concepts, Contexts and Cases*, Englewood Cliffs, N.J.: Prentice-Hall, pp. 439-456.

Quinn, J.B. (1988b). Innovation and corporate strategy: managed chaos. In M.L. Tushman and W.L. Moore (eds), *Readings in the Management of Innovation*, Cambridge, Mass.: Ballinger, pp. 123-137.

Quinn, J.B. (1993). Managing the intelligent enterprise: knowledge and service-based strategies. *Planning Review*, Vol. 21 (5), pp. 13-16.

Rayner, B.C.P. (1991). A blueprint for competition. *Electronic Business*, Vol. 17 (6), pp. 44-48.

Reich, R.B. and Mankin, E.D. (1986). Joint ventures with Japan give away our future. *Harvard Business Review*, March-April, pp. 78-86.

Ricardo, D. (1891). *Principles of Political Economy and Taxation*. London: G. Bell.

Roberts, E. (1995). Benchmarking the strategic management of technology, II. *Research-Technology Management*, Vol. 38 (2), pp. 18-26.

Roberts, E.B. and Berry, C.A. (1988). Entering new businesses: selecting strategies for success. In M.L. Tushman, and W.L. Moore (eds), *Readings in the Management of Innovation,* 2nd edn. Cambridge, Mass.: Ballinger.

Roberts, H.J.E. (1993). *Accountability and Responsibility: The Influence of Organization Design on Management Accounting*. Maastricht: University Press of Maastricht.

Robinson, W.T. (1988). Sources of market pioneer advantages: the case of industrial goods industries. *Journal of Marketing Research*, Vol. 25 (September), pp. 87-94.

Robinson, W.T. and Fornell, C. (1985). The sources of market pioneer advantages in consumer goods industries. *Journal of Marketing Research*, Vol. 22 (August), pp. 297-304.

Robinson, W.T., Fornell, C. and Sullivan, M. (1992). Are market pioneers intrinsically stronger than later entrants? *Strategic Management Journal.* Vol. 13 (8), pp. 609-624.

Rogers, E.M. (1983). *Diffusion of Innovation*. New York: Free Press.

Rooij, A.H. de and Cobbenhagen, J.W.C.M. (1991) Innovatie bij DSM: een praktijk-verhaal. *Bedrijfskunde*. Vol. 1991, no. 2, pp. 168-177.

Roos, J. and Krogh, G. von (1992). Figuring out your competence configuration. *European Management Journal*, December, Vol. 10 (4), pp. 422-427.

Roth, M.S. and Amoroso, W.P. (1993). Linking core competencies to customer needs: strategic marketing of health care services. *Journal of Health Care Marketing*, Vol. 13 (2), pp. 49-54.

Rothwell, R. (1972). *Factors for Success in Industrial Innovation. Project SAPPHO: A Comparative Study of Success and Failure in Industrial Innovation*. Brighton, Sussex: Social Policy Research Unit.

Rothwell, R. (1974). SAPPHO updated: Project SAPPHO phase II. *Research Policy*, Vol. 3, pp. 30-38.

Rothwell, R. (1976). *Innovation in Textile Machinery: Some Significant Factors in Success and Failure*. SPRU Occasional Paper Series, no. 2. Brighton, Sussex: University of Sussex.

Rothwell, R. (1977). The characteristics of successful innovators and technically progressive firms. *R&D Management*, Vol. 7 (3), 191-206.

Rothwell, R. et al. (1974). The Hungarian SAPPHO: some comments and comparision. *Research Policy,* Vol. 3, pp. 258-291.

Rotter, J. (1966). Generalized expectancies for internal versus external control of reinforcement. *Psychological Monographs*, Vol. 80 (609), pp. 1-28.

Roussel, P.A., Saad, K.N. and Erickson, T.J. (1991). *Third Generation R&D*. Boston, Mass.: HBS Press.

Rozeboom, W.W. (1966). *Foundations of the Theory of Prediction.* Homewood, Ill.: Dorsey Press.

Rubenstein, A.H., Chakrabarti, A.K., O'keefe, R.D., Souder, W.E. and Young, H.C. (1976). Factors influencing success at the project level. *Research Management,* Vol. 16 (May), pp. 15-20.

Rubin, D.B. (1973). The use of matched sampling and regression adjustment to remove bias in observational studies. *Biometrics,* Vol. 29, pp. 185-203.

Rudestam, K.E. and Newton, R.R. (1993). *Surviving your Dissertation.* London: Sage.

Rumelt, R.P. (1991). How much does industry matter? *Strategic Management Journal,* Vol. 12 (3), pp. 167-185.

Sandersons, S. and Uzumeri, V. (1990). *Strategies for New Product Development and Renewal: Design Based Incrementalism.* Center for Science and Technology Working paper. Troy, N.Y.: Rensselaer Polytechnic Institute.

Saren, M.A. (1984). A classification and review of models of the intra-firm innovation process. *R&D Management,* Vol. 14 (1), pp. 11-24.

Schein, E.H. (1985). *Organizational Culture and Leadership.* San Francisco: Jossey-Bass.

Schein, E.H. (1988). *Innovative Cultures and Organizations.* MIT/Sloan School of Management, Research Memorandum 90s, 88-064. Boston, Mass.: MIT.

Schein, E.H. (1989). Corporate culture is the key to creativity. *Business,* May, pp. 73-75.

Scherer, F.M. (1975). Firm size, market structure, opportunity and the output of patent invention. *American Economic Review,* Vol. 55 pp. 1097-1125.

Scherer, F.M. (1982). Inter-industry technology flows in the United States. *Research Policy,* Vol. 11, pp. 227-245.

Schoemaker, E.G. (1983). *The Diffusion of Innovation.* New York: Free Press.

Schoemaker, P.J.H. (1992). How to link strategic vision to core capabilities. *Sloan Management Review,* Fall, pp. 67-81.

Schön, D.A. (1983). *The Reflective Practitioner: How Professionals Think in Action.* New York: Basic Books.

Schön, D.A. (1984). *The Crisis of Professional Knowledge and the Pursuit of an Epistemology of Practice.* Harvard Business School 75th Anniversary Colloquium on Teaching the Case Method.

Schreuder, H., Cayseele, P. van, Jaspers, P. and De Graaf, B. (1991). Successful bear-fighting strategies. *Strategic Management Journal,* Vol. 12, pp. 523-534.

Schroeder, R. and Ven, A. van de, Scudder, G. and Polly, A. (1986). Managing innovation and change processes: findings from the Minnesota innovation research program. *Agribusiness,* Vol. 2 (4), pp. 501-523.

Schumpeter, J.A. (1934). *Theory of Economic Development.* Cambridge, Mass.: Harvard University Press.

Sclesselman, J.J. (1982). *Case-control Studies: Design, Conduct, Analysis.* Oxford: Oxford University Press.

SCP (1992). *Sociaal en cultureel rapport 1992.* The Hague: VUGA.

Seashore, S.E. (1987). Surveys in organizations. In J.W. Lorsch (ed.), *Handbook of Organizational Behavior,* Englewood Cliffs, N.J.: Prentice-Hall.

Selznick, P. (1949). *TVA and the Grass Roots*. Berkeley, Cal.: University of California Press.

Selznick, P. (1952). *The Organizational Weapon*. New York: McGraw-Hill.

Selznick, P. (1957). *Leadership in Administration: A Sociological Interpretation*. New York: Harper and Row.

Senge, P. (1990). *The Fifth Discipline: The Art and Practice of the Organisational Learning*. Maidenhead, Berks.: McGraw-Hill.

Shaw, B. (1985). The role of the interaction between the user and the manufacturer in medical equipment innovation. *R&D Management*, Vol. 15 (4), pp. 283-292.

Shrivastava, P. (1983). A typology of organizational learning systems. *Journal of Management Studies*, Vol. 20 (1), pp. 7-28.

Shrivastava, P. and Grant, J. (1985). Models of strategic decision making. *Strategic Management Journal*, Vol. 6, pp. 97-113.

Simon, H.A. (1991). Bounded rationality and organizational learning. *Organization Science*, Vol. 2, pp. 125-134.

Simonse, A., Kerkhoff, W. and Rip, A. (1989). *Technology Assessment in Ondernemingen*. Deventer: Kluwer Bedrijfswetenschappen.

Sitter, L.U. de (1987). *Het flexibele bedrijf*. Deventer: Kluwer Bedrijfswetenschappen.

Sitter, L.U. de (1989). *Integrale produktievernieuwing: Sociale en economische achtergronden van het TAO-programma*. Maastricht: MERIT.

Sitter, L.U. de (1994). *Synergetisch produceren*. Assen: Van Gorcum.

Sitter, L.U., Hertog, J.F. den and Dankbaar, B. (1997). From complex organisations with simple jobs to simple organisations with complex jobs. *Human Relations*, Vol. 50 (5), pp. 497-534.

Skinner, W. (1974). The focused factory. *Harvard Business Review*, Vol. 52 (3), pp. 113-121.

Sluijs, E. van and Hertog, J.F. den (1993). Praktijkverkenning Personeelswetenschappen. In *Deelrapporten Verkenning Personeelswetenschappen*, Tilburg: MERIT, Ou, WORC, Vakgroep Personeelwetenschappen, IVA.

Sluijs, E. van, Assen, A. van and Hertog, J.F. den (1991). Personnel management and organizational change: a sociotechnical perspective. *European Work and Organizational Psychologist*, Vol. 1 (1), pp. 27-51.

Smith, P.C., Kendall, L.M. and Hulin, C.L. (1969). *The Measurement of Satisfaction in Work and Retirement*. Chicago: Rand McNally.

Sneep, A. (1994). Innovation management in the Dutch agro/food industry. European agriculture on the move: the innovation era. *Tinbergen Institute PhD Research Bulletin*, Vol. 6 (3), pp. 17-28.

Snow, C.C. and Hambrick, D.C. (1980). Measuring organizational strategies: some theoretical and methodological problems. *Academy of Management Review*, Vol. 5, pp. 527-538.

Snow, C.C. and Hrebiniak, L.G. (1980). Strategy, distinctive competence, and organizational performance. *Administrative Science Quarterly*, Vol. 25, June, pp. 317-336.

Sommerlatte, T. (1986). 1000 Unternehmen antworten: die Innovationswelle komt. In A.D. Little (ed.), *Management der Geschäfte von Morgen*, Wiesbaden: Gabler, pp. 17-25.

Souder, W.E. (1987). *Managing New Product Innovations.* New York: Lexington.

Souder, W.E. (1988). Managing relations between R&D and marketing in new product development projects. *Journal of Product Innovation Management*, Vol. 5 (1), pp. 6-19.

Spendolini, M.J. (1992). *The Benchmarking Book.* New York: Amacom.

Stacey, R.D. (1990). *Dynamic Strategic Management for the 1990s.* London: Kogan Page.

Stacey, R.D. (1992). *Managing the Unknowable: Strategic Boundaries Between Order and Chaos in Organizations.* San Francisco: Jossey Bass Publishers.

Stacey, R.D. (1993). *Strategic Management and Organisational Dynamics.* London: Pitman Publishing.

Stalk, G. (1988). Time, the next source of competitive advantage. *Harvard Business Review*, July-August, pp. 41-51.

Stalk, G. (1992). Time-based competition and beyond: competing on capabilities. *Planning Review* (Special Issue), September-October, pp. 27-29.

Stata, R. (1989). Organizational learning: the key to management innovation. *Sloan Management Review*, Vol. 30 (3), pp. 63-74.

Stata, R. (1992). Management innovation. *Executive Excellence*, Vol. 9 (6), pp. 8-9.

Stevens, S.S. (1946). On the theory of scales of measurement. *Science*, Vol. 103, pp. 677-680.

Stevens, S.S. (1951). Mathematics, measurement, and psychophysics. In S.S. Stevens (ed.), *Handbook of Experimental Psychology*, New York: John Wiley.

Stewart, T.A. (1991). Brainpower. *Fortune*, June, pp. 44-60.

Stewart, T.A. (1994). Intellectual capital. *Fortune*, October, pp. 68-74.

Stobauch, R. (1988). *Innovation and Competition.* Boston, Mass.: Harvard Business School Press.

Storey, D.J. (1994). *Understanding the Small Business Sector.* London: Routledge.

Stouffer, S.A. (1949). *The American Soldier.* New York: Princeton University Press.

Strauss, A. and Corbin, J. (1990). *Basics of Qualitative Research: Grounded Theory, Procedures and Techniques.* Newbury Park, Cal.: Sage.

Sudman, S. (1976). *Applied Sampling.* New York: Academic Press.

Swann, G.M.P. (ed.). (1993). *New Technologies and the Firm.* London: Routledge.

Sweeney, E.P. and Hardaker, G. (1994). The importance of organizational and national culture. *European Business Review*, Vol. 94 (5), pp. 3-14.

Swieringa, J. and Wierdsma A.F.M. (1990). *Op Weg naar een lerende organisatie.* Groningen: Wolters Noordhoff.

Takeuchi, H. and Nonaka I. (1986). The new new product development game. *Harvard Business Review*, January-February, pp. 137-146.

Tannenbaum, A.S. (1968). *Control in Organizations.* New York: McGraw-Hill.

Taylor, W. (1990). The business of innovation: an interview with Paul Cook. *Harvard Business Review*, March-April, pp. 97-106.

Teece, D.J. (1982). Towards an economic theory of the multiproduct firm. *Journal of Economic Behaviour and Organisation*, Vol. 3, pp. 39-63.

Teece, D. and Pisano, G. (1994). The dynamic capabilities of firms: an introduction. *Industrial and Corporate Change*, Vol. 3 (3), pp. 537-556.

Teixeira, D. and Ziskin, J. (1993). Achieving quality with customer in mind. *Bankers Magazine*, Vol. 176 (1), pp. 29-35.

Thamhain, H.J. (1990). Managing technologically innovative, team efforts toward new product success. *Journal of Product Innovation Management*, Vol. 7, pp. 5-18.

Theeuwen, H.W.A. and Polastro, E.T. [1989]. Fijnchemie uitdaging in de volgende eeuw. *Chemisch Weekblad*, Vol. 42, pp. 408-411.

Thurstone, L.L. (1927). The method of paired comparison for social values. *Journal of Abnormal and Social Psychology*, Vol. 21, pp. 384-400.

Thurstone, L.L. (1929). Theory of attitude measurement. *Psychological Bulletin*, Vol. 36, pp. 222-241.

Toffler, A. (1980). *The Third Wave*. New York: William Morrow.

Tollison, R.D. (1982). Rent seeking: a survey. *Kyklos*, Vol. 35, pp. 575-602.

Trist, E.L. (1979). The evolution of sociotechnical systems as a conceptual framework and as an action research program. in A.H. van de Ven and W.F. Joyce (eds), *Perspectives on Organization Design and Behavior*, New York: John Wiley, pp. 19-75.

Turner, D. and Crawford, M. (1992). Managing current and future competitive performance: the role of competence. Paper presented at the workshop 'Competence Based Competition: Towards the Roots of Sustainable Competitive Advantage', Genk.

Turner, I. (1996). Strategy, complexity and uncertainty. Henley Management College. http://www.henleymc.ac.uk/products/ft4.htm

Tushman, M.L. and Anderson, P. (1986). Technological discontinuities and organisational environments. *Administrative Science Quarterly*, Vol. 31, pp. 439-465.

Tushman, M.L. and Moore, W.L. (1982). *Readings in the Management of Innovation*. Boston, Mass.: Pitman.

Tushman, M.L. and Nelson, R.R. (1990). Introduction: technology, organizations and innovation. *Administrative Science Quarterly*, Vol. 35 (1), pp. 1-8.

Tversky, A. and Kahnemann, D. (1974). Judgment under uncertainty: heuristics and biases. *Science*, Vol. 185, pp. 1124-1131.

Twiss, B. (1992). *Managing Technological Innovation*, 4th edn. London: Pitman.

Ulrich, D. and Lake, D. (1990). *Organizational Capability: Competing from the Inside Out*. New York: John Wiley.

UNESCO (1973). *Questionnaire of the International Comparative Study on the Organization of Research Units*. Paris: UNESCO.

Urban, G.L. and Hippel, E. von (1989). Lead user analyses for the development of new industrial products. *Management Science*, Vol. 34 (5), pp. 569-582.

Urban, G.L., Carter, R., Gaskin, S. and Mucha, Z. (1986). Market share rewards to pioneering brands: an empirical analysis and strategic implications. *Management Science*, Vol. 6 (June), pp. 645-659.

Uttal, B. (1987). Speeding new ideas to market. *Fortune*, Vol. 115 (5), pp. 54-57.

Vaill, P.B. (1990). Aanhoudend wild water. *Filosofie in bedrijf*, Vol. 2 (1), pp. 2-11.

Vall, M. van der (1975). Utilization and methodology of applied social research: four complementary models. *Journal of Applied Behavioral Science*, Vol. 11 (1), pp. 14-38.

Vall, M. van der (1980). *Sociaal beleidsonderzoek. Een professioneel paradigma.* Alphen aan de Rijn: Samsom.

Van Maanen, J. and Barley, S.R. (1984). Occupational communities: culture and control in organizations. In: B. Staw and L.L. Cummings (eds), *Research in Organizational Behavior,* Vol. 6, Greenwich, Conn.: JAI Press.

Varadarajan, P.R. and Ramanujam, V. (1990). The corporate performance conundrum: A synthesis of contemporary views and an extension. *Journal of Management Studies*, Vol. 27 (5), pp. 463-483.

Vaus, D.A. de (1991). *Surveys in Social Research*, 3rd edn. London: UCL Press/Allen and Unwin.

Ven, A.H. van de (1986). Central problems in the management of innovation. *Management Science*, Vol. 32, pp. 590-608.

Ven, A.H., van de, Angle, H.L. and Scott Poole, M. (1989). *Research on the Management of Innovation: The Minnesota Studies.* New York: Harper and Row.

Ventriss, C. (1990). Organizational theory and structure: an analysis of three perspectives. *International Journal of Public Administration*, Vol. 13 (6), pp. 777-798.

Ventriss, C. and Luke, J. (1988). Organizational learning and public policy: towards a substantive perspective. *American Review of Public Administration*, Vol. 18 (4), pp. 337-357.

Volberda, H.W. (1990). Een flexibele organisatie als voorwaarde voor innovatie. *M&O*, Vol. 44 (3), pp. 215-242.

Von Hippel, E. (1976). Users as innovators. *Technology Review*, Vol. 5, pp. 212-239.

Von Hippel, E. (1978). A customer-active paradigm for industrial product idea generation. *Research Policy*, Vol. 7, pp. 240-266.

Von Hippel, E. (1982). Appropriability of innovation benefits as a predictor of the source of innovation. *Research Policy*, Vol. 11, pp. 95-116.

Von Hippel, E. (1986). Lead users: A source of novel product concepts. *Management Science*, Vol. 32 (7), pp. 791-805.

Voss, C.A. (1988). Implementation: a key issue in manufacturing technology. The need for a field for study. *Research Policy*, Vol. 17, pp. 55-63.

Vught, F.A. van (1987). Pitfalls of forecasting. *Futures*, April, pp. 184-196. Also published in R.A. Burgelman and L.R. Sayles (eds), *Inside Corporate Innovation: Strategy, Structure, and Managerial Skills.* New York: Free Press, 1986.

Warthouse, B. and Louie, D. (1989) An overview of financial activities of the specialty chemicals industry. In *Specialty Chemicals: Strategies for success*, Vol. 1. Menlo Park: SRI.

Weber, M. (1947). *The Theory of Social and Economic Organization,* tr. Henderson and Parsons. New York: Oxford University Press, pp. 329-334.

Weinrauch, J.D. and Anderson, R. (1982). Conflicts between engineering and marketing units. *Industrial Marketing Management*, Vol. 11, pp. 291-301.

Weiss, A.R. and Birnbaum P.H. (1989). Technological infrastructure and the implementation of technological strategies. *Management Science*, Vol. 35 (8), pp. 1014-1026.

Weiss, R.S. and Rein, M. (1970). The evaluation of broad-aim programs: experimental design, its difficulties and an alternative. *Administrative Science Quarterly,* Vol. 15 (3), pp. 97-109.

Wernerfelt, B. (1984). A resource-based view of the firm. *Strategic Management Journal*, Vol. 5, pp. 171-180.

Wernerfelt, B. (1989). From critical resources to corporate strategy. *Journal of General Management*, Vol. 5, pp. 171-180.

Whitney, D.E. (1988). Produceren volgens ontwerp. *Harvard Holland Review*, Winter, Vol. 17, pp. 41-49.

Wilkins, A. (1989). *Developing Corporate Character: How to Successfully Change an Organization without Destroying It.* San Francisco, Cal.: Jossey-Bass.

Willems and Van den Wildenberg B.V. (1993). *New Business Development op Basis van Technische R&D.* The Hague: Willems and Van den Wildenberg B.V.

Winsemius, P. [1988]. Hand in hand: twee culturen in één bedrijf. In *Jaarverslag VNCI 1987, Vereniging van de Nederlandse Chemische Industrie.* Leidschendam: VNCI.

Wit, B. de (1994). Resource-based strategy: een nieuw paradigma. In A. Witteveen (ed.), *Top 20 Trends in Strategisch Management.* Amsterdam: Management Press, pp. 131-139.

Witteveen, A. (1994). Intentie en competentie. In A. Witteveen (ed.), *Top 20 Trends in Strategisch Management.* Amsterdam: Management Press, pp. 120-123.

Womack, J., Jones D. and Roos, D. (1990). *The Machine that Changed the World.* New York: Rawson Associates, Maxwell Macmillan Int.

Wood, S. (1990). Tacit skills, the Japanese model and new technology. *Applied Psychology*, Vol. 39 (2), pp. 169-190.

Woodward, J. (1965). *Industrial Organization: Theory and Practice.* Oxford: Oxford University Press.

Yin, R.K. (1984). *Case Study Research. Design and Methods.* London: Sage.

Yin, R.K. (1990). *Case Study Research. Design and Methods,* rev.edn. Beverly Hills, Cal.: Sage.

Zaltman, G., Duncan, R. and Holbeck, J. (1973). *Innovation and Organization.* New York: John Wiley.

Zwan, A. van der, Bijvoet, L.C.L. and Jaspers, P.F.M. (1987*). Koplopers en achterblijvers in de bedrijven wereld.* The Hague: Nationale Investeringsbank.

Index

Curriculum Vitae

Dr. Jan Cobbenhagen is a senior researcher at MERIT (Maastricht Economic Research Institute on Innovation and Technology). He is program leader of the research program 'Innovation Management in Firms and Regions'. From 1990 to 1999, he was part time 'Maître de Conférences' at the University of Liège and from 1991 to 1995 he was part time visiting professor at the Universidad Carlos III, Madrid. As of February 2000 Dr. Cobbenhagen is managing director of M3 New Business Creation.

Date Due

McK DUE	FEB 1 3 2002	
McK DUE	MAY 0 2 '03	
McK DUE	FEB 1 6 2004	
MCK RTD	FEB 1 2 2004	